DEATH FROM THE SKIES!

DEATH FROM THE SKIES!

These Are the Ways the World Will End . . .

PHILIP PLAIT, Ph.D.

VIKING

VIKING
Published by the Penguin Group
Penguin Group (USA) Inc., 375 Hudson Street, New York, New York 10014, U.S.A. • Penguin Group (Canada), 90 Eglinton Avenue East, Suite 700, Toronto, Ontario, Canada M4P 2Y3 (a division of Pearson Penguin Canada Inc.) • Penguin Books Ltd, 80 Strand, London WC2R 0RL, England • Penguin Ireland, 25 St. Stephen's Green, Dublin 2, Ireland (a division of Penguin Books Ltd) • Penguin Books Australia Ltd, 250 Camberwell Road, Camberwell, Victoria 3124, Australia (a division of Pearson Australia Group Pty Ltd) • Penguin Books India Pvt Ltd, 11 Community Centre, Panchsheel Park, New Delhi–110 017, India • Penguin Group (NZ), 67 Apollo Drive, Rosedale, North Shore 0632, New Zealand (a division of Pearson New Zealand Ltd) • Penguin Books (South Africa) (Pty) Ltd, 24 Sturdee Avenue, Rosebank, Johannesburg 2196, South Africa

Penguin Books Ltd, Registered Offices: 80 Strand, London WC2R 0RL, England

First published in 2008 by Viking Penguin, a member of Penguin Group (USA) Inc.

10 9 8 7 6 5 4 3 2 1

Library of Congress Cataloging-in-Publication Data

Plait, Philip C.
Death from the skies! : these are the ways the world will end / Philip Plait.
p. cm.
Includes index.
ISBN 978-0-670-01997-7
1. End of the world (Astronomy) I. Title.
QB638.8.P53 2008
520—dc22 2008022943

Printed in the United States of America
Set in Minion with Berthold City
Designed by Daniel Lagin

Contents

DEATH FROM THE SKIES!

Introduction

Introduction

THE UNIVERSE IS TRYING TO KILL YOU.

It's nothing personal. It's trying to kill me too. It's trying to kill *everybody.*

And it doesn't even have to try very hard.

The Universe is an incredibly hostile place for life. Virtually all of it is a vacuum, so that's bad right from the start. Of the extremely few places that aren't hard vacuum, most are too hot for chemical reactions to do very well—molecules get blasted apart before they can even properly form. Of the places that aren't too hot, most are too cold—reactions happen too slowly to get interesting things to occur in the first place.

And in the very few places that aren't in a vacuum, too hot, or too cold—and we really know of only one: Earth—all manner of dangers are lurking about. Volcanoes blast megatons of noxious chemicals into the air, spew lava for miles, and cause vast earthquakes. Tsunamis rewrite huge sections of coastlines. Ice ages come and go; mountains pop up and change the global weather system; whole continents get subducted into the glowing rock of the mantle.

And those are just the *local* problems. Earth still sits in the incredibly hostile environment of space, and all kinds of disasters might befall us from there—literally.

But then, that's what this whole book is about.

Other job markets may lay claim to the title, but astronomy is actually the world's oldest profession: early agricultural civilizations needed to know when to plant their crops and when to harvest, and the changing skies gave them their clues. The appearance of a certain constellation at a certain time was as good as marking a calendar.

The Sun was worshipped, and the Moon. This evolved into the belief that all gods lived in the sky. Studying the sky was tantamount to worship.

Astrology arose, attempting (and failing, I'll note) to tie people's lives to the sky. With the invention of the telescope, and cameras to fit onto them, the sky was found to be more than just a reflection of our superstitions. It became a canvas for some of the finest artwork humans have ever seen. Its vista boasted dusty, ribboned nebulae; sweeping, majestic galaxies; layered, mottled planets. It was a thing of beauty.

Astronomy became even more of a science, guided by principles of physics, mathematics, and chemistry. It blossomed into a field in its own right, and spawned many more.

And during all this time, through all those millennia, it was always assumed that the Universe was a place *designed* for humans. Everything must be just so to support us, and clearly this was the way things were meant to be.

Hubris. Silly, silly hubris.

Because the Universe is a violent place. Stars explode. Stars like the Sun can die in milder events, but die just the same. Whole galaxies collide, igniting fireworks on a cosmic scale. Asteroids impact other planets; could they hit us?

When we launched telescopes into space, we equipped them with cameras that could detect ultraviolet light, X-rays, super-high-energy gamma rays. We saw a Universe that seemed actively trying to destroy us. Exploding stars are phenomenally dangerous, blasting out vast amounts of killing force and energy. Black holes are everywhere, lurking throughout the galaxy, devouring anything that wanders too close. Flashes of high-energy light from distant points in the Universe whisper of powers terrible and gross, enough to fry entire solar systems that get in the way.

The Earth seemed to be the center of the Universe for much of mankind's history. Now, however, the Earth seems fragile and impossibly small, a remote speck of dust lost in a Universe of frightening size and age.

In reality, the Universe cares not at all if we live or die. If a human were magically transported to any random spot in the cosmos, within seconds he'd die 99.9999999999999 percent of the time. At best.

Yet, despite all that (and quite a bit more still unsaid), *here we are.* Billions of years in, countless times around the Sun, sitting at the crosshairs of dozens of cosmic weapons . . . our planet endures. Life not only survives, it *thrives.* Numerous setbacks have occurred, for sure, but life itself continues. As small and fragile and soft as humans are, we've managed so far.

Of course, we haven't yet seen everything the Universe can unleash on us. A single asteroid impact could take out half of humanity without even breaking a sweat. A solar flare could wipe out our economy in seconds. And a nearby gamma-ray burst . . . well, that's bad too. *Very* bad.

I love astronomy. I've devoted my entire life to it, to telling others about it, to writing about it. Astronomy is awe-inspiring, it's beautiful, it provides philosophical perspective and holds the secret answers to so many of our most profound questions.

And I have no doubt whatsoever that astronomy can kill us. Well, no, that's poorly phrased. Let me say that astronomical *events* can kill us. In some cases, our study of astronomy can actually save us. In others it provides us with information about what might kill us, though unfortunately without giving us any ideas of how to stop it.

And in many of those cases, there isn't much we could do anyway.

This book is about those events. An asteroid impact killed the dinosaurs, and another rock may be out there with our name on it. The Sun seems constant, but is capable of doing serious damage. Supernovae wreak horrific destruction on epic scales. We'll visit these scenarios, and many more. We'll explore just what would happen if a black hole decided to pay us a visit, and what we can do as we ride out the eventual and inevitable death of the Sun, six billion years from now.

We'll even have a front row seat as the Universe ages countless years, and see what happens at the end of time.

These topics have been written about before, of course. They are certainly the subjects of many breathless documentaries on TV. Most of these fall quite a bit short of reality; they exaggerate the damage, or underestimate it. They play up minor aspects and ignore major ones. They rarely, if ever, talk about the actual likelihood of such an event.

That last point is a critical one. Over the years I've written about astronomical disasters both real and imagined, and many people get honestly scared about them. Anytime an asteroid is predicted to pass by the Earth they envision an apocalyptic scenario, fueled by reporters who play up the danger without mentioning that the odds of our getting hit are less than the chance of winning the typical lottery. I've spent far too much time assuaging people's fears, both rational and otherwise.

In this book, I won't hold back. The reality of a nearby gamma-ray burst puts the sweatiest fundamentalist religion's Armageddon prose to shame, dwarfing it to mundanity. I will go over, in loving detail, the Earth's atmosphere ripped away, the oceans boiled, and all life sterilized down to the base of the crust.

But during all that, I will remind you that there is no star nearby capable of creating such a burst; and even if there were, the odds of it going off anytime soon are tiny; and even if it did, the odds of it being aimed our way are tinier yet.

But it's still fun to think about "What if . . . ?"

While you're reading this book, you may feel like you're watching a horror movie at the theater: it's fun, jolting, and maybe even terrifying.

During the scary parts you may want to turn away, or hide your eyes, or spill your popcorn, but I'll make sure the actual facts of the case are there to calm you down a bit afterward.

Of course (he says, chuckling low and with evil intent), there *is* a big difference: eventually the movie is over, you leave the theater, and you laugh at the scary ride.

You can't do that in the real world. There *are* dangers out there, and we can't avert our eyes from them. But as you read this book (I hope with your eyes open) you'll learn just what the dangers are and, more important, what they aren't. What horror movie is still scary once the lights are on?

And you always have to keep in mind that *we're still here.* The Universe is a dangerous place, but again, we've gotten this far. We may just make it a while longer.

Or we may not. I have to be honest. The Universe is vast beyond imagining, and wields mighty forces. For nearly all the events depicted here, it's not a matter of *if,* it's a matter of *when.*

CHAPTER 1

Target Earth: Asteroid and Comet Impacts

THE ALARM WENT OFF AT 6:52 A.M. AS IT DID EVERY *morning. Groggily, Mark slapped it off, then stumbled wearily to the bathroom. He splashed a bit of water on his face to accelerate the waking process, then began to brush his teeth.*

Seeing that it was already a clear, warm day, he peered out the bathroom window to take in the scene as he brushed. The trees were covered in leaves, and flowers were in full bloom. The trees cast long shadows as the Sun slowly rose in the sky.

When he finished brushing, Mark noticed the odd silence. That's funny, *he thought.* Why aren't the birds chirping? *From the corner of his eye he saw movement. Maybe it was an animal in the yard that had spooked the birds . . .*

Stepping up to the window, he stood on tiptoe to look around the yard. What the— *Every tree was casting* two *distinct shadows. His morning routine now forgotten, Mark watched in amazement as, for every tree, one of the shadows appeared to be moving, circling around the base of the tree like fast-motion video of a sundial. Nose pressed to the window, he looked up into the sky, straining to see what could be causing this strange display.*

Suddenly, from under the eave, it appeared as if the Sun itself were streaking across the sky. Dazzled, Mark's eyes took a moment to adjust, but it was still not clear what he was seeing. There was a disk of intense white light moving across the sky, faster than an airplane. Could it be a meteor?

It appeared to descend slowly to the horizon as he watched. Then, in the blink of an eye, there was a soundless but all-encompassing flash, so bright his eyes watered. He winced in pain. When he was able to look again, the small bright disk was gone, replaced by a much larger smear of light, fanning up from the horizon. The heat from the thing was palpable, even through the window. It was like standing near a fireplace. As the smudge in the sky expanded, Mark noticed something even odder: did the tops of the trees look funny? Was that smoke *rising from them . . . ?*

The heat became intense. It began to dawn on Mark that he might be in trouble. As he stood there wondering what to do, a sudden and sharp earthquake jolted the house, knocking him to the floor. It was over quickly, and as he stood up, dazed, he felt the heat more strongly than before as it poured through his now-broken bathroom window. He thought the worst was over, but what he didn't know was that a wave of pure sound and fury tearing through the atmosphere was pounding toward him at 700 miles per hour.

Too late, he saw the face of the shock wave bearing down on him like a tsunami ten miles high. A mighty thunderclap swept over his burning house, pulverizing it to dust with Mark still inside, and the time for decisions was over.

Everything under this wave of sound was stomped flat. Trees that were ablaze a moment before from the heat of the explosion were snuffed out, then torn into millions of splinters. The expanding ring of pressure, already dozens of miles across, screamed past the location of Mark's disintegrated house and continued moving, greedily consuming buildings, trees, cars, people.

Before it was over, the shock wave circled the Earth twice. Seismographs from around the globe registered the event as an earthquake of enormous scale, but no one paid attention to the scientific data for long. They were too busy struggling to survive.

METEORS AND METEOROIDS AND METEORITES, OH MY!

The Earth sits in a cosmic shooting gallery, and the Universe has us dead in its crosshairs.

Consider this: the Earth is pummeled by twenty to forty *tons* of meteors every single day. Over the course of a year, that's easily enough to fill a six-story office building with cosmic junk.

While that sounds like a lot, it's really only a pittance compared to the size of the Earth, which is about a quintillion—a million million *million*—times bigger. But space is swarming with debris, and the Earth is constantly plowing through it.

The vast majority of this material is detritus, tiny bits of rock that burn up readily in our atmosphere. When you go out on a dark, clear night, you see these as "shooting stars," what astronomers call *meteors.* You might be surprised to find out that even the brightest ones you're likely to see are caused by tiny bits of fluff called *meteoroids,* no bigger than a grain of salt. Something as small as a pea would make a fantastically bright meteor—I once saw one that was so bright it lit up the sky and even left an afterimage on my eye. I stood transfixed for the two or three seconds it took to flash across the sky, but was just as shocked when I later calculated that the rock itself was probably no bigger than a grapefruit.

How can something so small get so bright? There are two factors to consider. You may be familiar with the first: compressing air heats it up. Think about how warm a bicycle pump gets after you use it—when the air is squeezed inside the pump, it gets hot and transfers that heat to the metal. You can actually burn yourself using a pump if you're not careful. The more a gas is compressed, the hotter it gets. The second factor is the fantastic speed at which meteoroids travel. Most of them hit us at ten to twenty miles per second, and some come roaring in as fast as sixty miles per second! This is far, far faster than even a rifle bullet.

When something moving that rapidly enters our atmosphere, its velocity is translated into energy, which in turn is transferred to the air around it. As it screams through the upper atmosphere, a meteoroid rams the air violently—a rock moving at Mach 50 is going to compress the air a *lot.* The air gets squeezed so quickly and at such high pressure that it heats up thousands of degrees and starts to glow.

As you can imagine, all that hot air is like a blast furnace. The meteoroid, traveling just a few inches behind that rammed air, feels that heat. It can't last long in those conditions, and if it's small it usually burns up in a matter of seconds. We see a bright glow, a streak across the sky that lasts for a moment or two, and then it's gone, adding its nearly insignificant mass to the Earth's.

To a stunned observer, a meteor looks like it's traveling just over his head, but in reality the action is occurring fifty or more miles above the ground. At that height the air is very thin, yet still thick enough to stop small, dense particles. But what if the particle is bigger than a pea, or a grape, or a watermelon? What if it's the size of, say, a couch, a car, a bus?

For a bigger object, things are very different. If it's a few yards across, instead of simply burning up, that chunk of space debris gets squeezed by the air pressure as if it's in a vise—the pressure can top out at over a thousand pounds per square inch at meteoric speeds. This pressure can flatten out the incoming object in a process called *pancaking* for obvious reasons. But a rock can only take so much of that before it crumbles and falls apart. Within seconds, instead of one big rock coming in, we now have hundreds or thousands of little ones, all still moving at velocities of several miles per second, and all dumping their energy into the air around them. They compress further, fracture, heat up, and so on . . . and within a fraction of a second we have a whole lot of rubble releasing a whole lot of heat all at once.

This is, by definition, an explosion.

So medium-sized meteoroids blow up in the atmosphere. Again, this usually happens fairly high up, depending on how tough the meteoroid is; ones made of metal can take more punishment and penetrate deeper into our atmosphere, but may still explode many miles above the Earth's surface. The energy involved is impressive: a rock only a meter across can explode with the force of hundreds of tons of TNT. In fact, military records indicate that such an explosion from an incoming chunk of rock is seen on average once a month!

Since meteoroids explode so high up in the atmosphere, you'd expect we'd be safe from things that size.

Well, not exactly. Under some conditions, the incoming rock may break up, but some chunks can survive. If the main mass slows enough before it explodes, then smaller fragments can slow even more without totally disintegrating. These can make it all the way to the ground. Metallic meteoroids have even more structural strength and can remain intact all the way down as well. If they do survive and impact the ground, they're called *meteorites.**

Small meteoroids that make it down to the ground usually aren't moving terribly fast when they impact. In fact, their initial velocity is completely nullified by our air, leaving them to fall at what is called *terminal velocity.* It's as if they were dropped off a tall building or from a balloon; they wind up impacting at maybe one or two hundred miles per hour. Scary, sure, but not *too* scary.

Still, you wouldn't want to get hit by a rock moving that fast. For comparison, they hit the ground faster than even a professional baseball player's pitch. In November 1954, a woman named Ann Hodges from Sylacauga, Alabama, was actually hit by meteorite. It was fairly small, about the size of a brick and weighing just over eight pounds. It punched through her roof, bounced off a wooden radio cabinet, and smacked into her where she was lying on the couch, taking a nap that

---

* One of the best ways to tick off an astronomer—and it can be fun sometimes just to see how he reacts—is to mix up the terms *meteor, meteoroid,* and *meteorite.* The very *best* way to tick off an astronomer is to call him an astrologer.

was rather rudely interrupted. Her hand and side were injured. She lived, but suffered one of the nastiest bruises in medical history.

This may be the earliest well-documented case of a meteorite damaging human property. But it wasn't the last. With the advent of the video camera, it was inevitable that more and more spectacular meteors would be recorded.

On October 2, 1992, a meteoroid the size of a school bus entered the Earth's atmosphere. It created a huge fireball as it traveled northeast across the United States, and was witnessed by thousands of people—by a happy coincidence, it was on a Friday night during football season, so many proud parents were already running their video cameras, yielding excellent footage of the meteor. The rock broke apart as it ripped its way across the sky, and one of the pieces, roughly the size of a football, fell onto the trunk of a young woman's car in Peekskill, New York. It left a hole in the back end of the car that looked, not surprisingly, exactly as if it had been caused by a rock dropped from a great height. One can imagine the difficulty the owner had getting her insurance company to pay for the damage.

These and other stories notwithstanding, in the end the Earth's surface is big, and most meteorites are small. The odds of anyone's getting

As it burned its way through the Earth's atmosphere, the Peekskill meteor was captured on dozens of home movie cameras. It broke into smaller chunks, one of which hit a woman's car.

SARAH EICHMILLER AND THE *ALTOONA (PA) MIRROR*

hit by one are really very small, and the odds of being killed by one are even smaller.

Still, *most* meteorites are small. Some aren't.

SHALLOW IMPACT

On June 30, 1908, the Earth and a smallish chunk of pretty weak rock found themselves at the same place at the same time.

The rock was probably seventy or so yards across. Its orbit intersected the Earth's, and over time it was inevitable that the two objects would both be located at that intersection point simultaneously.

It came in over Siberia, in a remote region near the Podkamennaya Tunguska River. On that day, it entered the Earth's atmosphere over Russia, traveling northwest. It plunged deeper into the air, and the increasing pressure put tremendous strain on the meteoroid. It broke apart, and each piece broke apart, and the cascade of rupture dumped a vast amount of energy into the air around it. The object exploded, releasing between three and twenty megatons of energy: the equivalent of three to twenty million tons of TNT, hundreds of times as much energy as the bomb dropped on Hiroshima thirty-seven years later.

The blast itself was seen by hundreds of witnesses (the Soviet Union even created a stamp based on what was seen), and the explosion registered on seismometers designed to detect earthquakes. People were knocked off their feet hundreds of miles away.

Despite the incredible event and the excitement it generated, a scientific expedition took years to mount. The region is unbelievably difficult to reach; in winter it's forbidding at best (we're talking Siberia after all), and in the summer the Tunguska region is a swamp, infested with mosquitoes. But eventually the site was reached, and what greeted those weary travelers had never been seen before.

As they approached the area of the explosion, the expedition members were shocked to see trees flattened like toothpicks for hundreds of square miles. Moreover, the trees were lying in parallel formations. Following the trail, the scientists came to a spot where the trees were all knocked over *radially,* like spokes on a bicycle wheel. Even weirder, the trees at ground zero were still standing, though totally denuded of branches and leaves. It's hard to imagine what they must have felt upon seeing such an eerie sight.

No blast crater was ever found, nor (yet) any definitive debris from the rock. It exploded several miles above the ground, and totally vaporized. The air blast created a shock wave that knocked down those trees. The trees at the center were still standing because the blast wave slammed straight down into them; it takes sideways force to knock trees down. Nuclear airburst blasts during weapons tests of the 1950s and 1960s replicated the same pattern.

While the remote location of the explosion made it hard to study, it also meant few people were killed. Had the explosion occurred over Moscow or London, millions would have died within minutes, making this a very serious threat indeed. Still, the immediate effect from the explosion was localized. Probably no one more than a few dozen miles away was hurt.

But then, not all impactors are only seventy yards across . . . and not all impacts are local.

PAIN IN THE ASTEROID

Sixty-five million years ago, the dinosaurs had a really bad day.

Actually, recent findings show they were having a bad couple of million years. There are indications that the Earth's climate had been changing, and many species were already in decline. However, there is overwhelming evidence that a great number of species indeed died practically overnight on a geological time scale. It's now a matter of scientific fact that this event was triggered by the impact of a six-mile-wide asteroid—and at that size, the word "meteoroid" is *seriously* inadequate.

It was certainly large enough to do the trick. The mind boggles to think of the devastation wrought when a rock bigger than Mount

Everest plummeted through the atmosphere and hit the Earth at ten miles per second. Imagine: when the surface of the asteroid contacted the ground, the far side was still sticking out above most of the Earth's atmosphere.

The exact energy of the impact is difficult to know, but it would have been hundreds of *millions* of megatons—far, far larger than the heftiest nuclear bomb ever detonated. In fact, even if you detonated every single nuclear weapon on Earth simultaneously, the explosion generated by the impact of the dinosaur killer would have been a million times more powerful . . . all concentrated in one spot.

The dinosaurs had a really, *really* bad day.

That massive impact set off a terrifying series of events, each of which brought destruction on an unimaginable scale.

As the asteroid plunged through the air, it would have created a huge shock wave, superheating the atmosphere for miles around it. As bright as the Sun, it would have set everything underneath it aflame even before it hit. And if anything did manage to survive that terrible heat, it would then have to withstand the force of a giant shock wave slamming into it as the asteroid tore through the air during its supersonic travel.

Being so large, the asteroid would hardly have slowed its flight or

Meteor Crater, in Arizona, formed in an impact about 50,000 years ago. The iron asteroid that gouged this crater out of the desert was only 50 yards across. The far rim wall is almost a mile away; note the people in the lower right for scale.

PHOTO BY THE AUTHOR

lost any mass at all before it slammed into the ground. Scientists now know that the impact point was just off the Yucatán Peninsula in Mexico. It impacted water—which isn't too surprising, as water covers 71 percent of the Earth's surface. A huge section of the Gulf of Mexico would have exploded into steam as the ferocious energy of the asteroid's motion was converted to heat upon impact. In the relatively shallow water there, the asteroid still would not have slowed much before hitting the continental shelf. Once it finally hit rock, the impacting mass would have stopped, and the remaining energy would have flash-converted to heat.

At this point, what was moments before a horrifying scenario turns into complete apocalypse as several events occur at once. Slamming into the Earth's crust, all those millions of megatons of energy exploded outward, sending molten rock and vaporized seawater upward and outward. The plume shot up miles into the sky, bright and hot as the Sun. The impact itself generated a huge ground shock wave, dwarfing any mere terrestrial seismic event and killing everything for hundreds of miles around the impact site.

Following the ground shock was an air shock, an epic sonic boom. Any creatures within a thousand miles that survived the initial impact were quite deaf once the thunderclap reached them.

But if they were anywhere near the Gulf of Mexico, they wouldn't have lasted long anyway. When the asteroid hit the water, it displaced vast amounts of the ocean, both because of the shock wave and through simple vaporization due to heat. What it created was a tsunami, but one on a huge scale.

In December 2004, an earthquake caused a tsunami a few yards high that moved slower than a car, yet killed a quarter of a million people. The tsunami generated by the asteroid impact was *hundreds* of yards high, and moved at *600 miles per hour.*

Within minutes a roaring mountain of billions of tons of seawater slammed into the Texas coast, scouring it clean of any life. The tsunami marched inland for miles, killing everything in its path with a fierce devastation no tornado, hurricane, or earthquake could ever hope to match.

And yet this impact still had more death to deal. When the asteroid hit, it punched a hole in the Earth right through the crust. The energy of the impact sent molten rock hurtling into the air at speeds of several miles per second. At those speeds, the debris would actually go up and out of the atmosphere on ballistic trajectories, like intercontinental missiles. As they fell back down, these ejecta would heat up and burn, replicating the original event on a miniature scale, but billions of times over. Flaming rock would fall from the sky like a cloudburst for thousands of miles around the impact point, igniting forest fires across the globe that would rage out of control and fill the Earth's atmosphere with thick black smoke.

Essentially, the whole planet caught fire.

Back at ground zero, the impact point itself would have been like nowhere else on Earth. A crater two hundred miles across and twenty miles deep was chewed into the crust, its temperature soaring to 6,000 degrees Fahrenheit. Inrushing water instantly vaporized, creating more devastation, if such a thing was even possible.

No place on Earth was left untouched. Fires blazed everywhere. As the world burned, the atmosphere darkened, letting very little sunlight through. Over time, the Earth cooled dramatically, eventually causing an ice age that would kill even the incredibly tough plants and animals that survived the initial onslaught.

Through sheer happenstance, the asteroid hit a spot on Earth that was rich in limestone and other minerals. The shock wave from the impact (and from ejected rock reentering the atmosphere) created nitrates from this material that then formed nitric acid in the air that rained down over the planet. Moreover, chlorine and other chemicals in the asteroid itself were released upon impact; catapulted into the upper atmosphere, they were sufficient to destroy the ozone layer thousands of times over. This killed not just plant life, but aquatic life as well. The food chain was disrupted at its most fundamental level on the whole planet, and when the fires finally died down, as much as 75 percent of all life on Earth was extinguished.

Eventually, the crater cooled, the fires went out, and the natural cycles of the Earth buried the evidence. Life remaining on the planet had it pretty tough for a long time, but with that much devastation there were many environmental niches to be filled. Life did as it always does—it found its path, and the Earth was repopulated.

Fast-forward sixty-five million years. Archaeologists digging through rock layers see a dramatic change in composition and color between two strata. Below this change are rocks and fossils from the Cretaceous period, and materials above it are from the Tertiary. This striking discontinuity, called the *K-T boundary* (unfortunately, the term *C-T* was already being used by archaeologists, so they had to settle for K-T), would be a mystery for decades, and not just among scientists: since it marked the end of the dinosaurs, it caught the public's imagination as well.

After years of investigation, the smoking gun turned up: a layer of iridium was found in the rock at the K-T boundary—it's an element rare on the surface of the Earth, but common in asteroids. Also, many areas on Earth have a layer of soot just above the K-T boundary, probably attesting to the global fires. Both pointed right to an impact from an asteroid. All that was needed to clinch the deal was the location of the crater.

It too was eventually found, centered just off the tip of the Yucatán Peninsula. You might think a huge crater would be easy to find, but in fact it's difficult. Millions of years of erosion eradicated many crater features. Plus, the crater itself, called Chicxulub,* is so big that it can only be seen easily from space. Amazingly, you could be standing in the middle of it and never know. It's so large but so difficult to measure that scientists are still arguing over its size and depth.

After all this—the global destruction, the extinction of countless species (including, of course, the dinosaurs, which had previously en-

---

* Pronounced "Cheek-shoo-lube."

joyed a pretty impressive two-hundred-million-year run), and an environmental impact that lasted for centuries—it might be worthwhile to note that the culprit, an asteroid six miles across, would be categorized by most astronomers as "small."

Much, much larger asteroids exist. Most never get near the Earth. But there are several of similar size that not only approach us, but have orbits that actually cross that of the Earth. For them, an impact is not a matter of *if.* It's a matter of *when.*

The dinosaurs had a very bad day, but our own day may yet come.

COSMIC WEAPONS DUMP

Where are all these rocks coming from?

The majority of asteroids in the solar system circle the Sun between the orbits of Mars and Jupiter in what's called the *asteroid belt,* or the *main belt.* There may be billions of them there, occupying several quintillion cubic miles of space in a volume resembling a flattened doughnut. Most are tiny, grains of dust, or pea-sized. The largest, Ceres, is about six hundred miles across, and was the first to be discovered. On January 1, 1801—the first day of the new century*—the Italian astronomer Giuseppe Piazzi found it while scanning the heavens. Knowing that astronomers had supposed that the gap between Mars and Jupiter might hide a small planet, and seeing that his object moved from night to night, Piazzi thought he had finally found it. However, within a few years several more objects were found in the same region of space. As a group, they were named *asteroids,* meaning starlike objects; they were too small and too far away to be anything more than points of light to the telescopes of the time.

The origin of the asteroids has been a mystery for a long time. At first, it was assumed that they were the rubble from a planet that existed between Mars and Jupiter that was somehow destroyed. Today, the

---

* Yes, it *was* the first day of the nineteenth century. Don't make me lecture you about there not being a year 0.

weight of accumulated evidence indicates that the asteroids are actually leftover detritus from the formation of the solar system. These scraps were never able to accumulate to form a major planet because of the powerful gravitational influence of Jupiter; the gravity of the solar system's largest planet accelerated the asteroids, increasing the speeds of their collisions. Instead of sticking together from low-speed collisions to form bigger objects, they hit at higher speeds, which shattered them.

Several hundred thousand asteroids are known today. Many have been discovered through dogged determination; astronomers huddled over their telescopes' eyepieces, watching the sky, night after night. Today, there are automated telescopes—robots, in a sense—that use preprogrammed patterns to scan the sky. The vast amounts of data generated are then analyzed by computer to look for moving objects. It's actually relatively rare these days for a human to find an asteroid.

While the majority of all known asteroids orbit the Sun in the main belt, not all of them do. Various processes, gravitational and otherwise, can change the shapes of the orbits of some main-belt asteroids over time. Their orbits can become more elliptical, dipping them closer in and farther out than the other asteroids in the main belt. Some cross Mars's orbit, and some cross the Earth's.

It's the latter we need to be concerned about.

The search for these Earth-crossers (called Near Earth Objects, and dangerous ones tagged as Potentially Hazardous Objects) is a multinational effort, but it's still somewhat small—fewer than two dozen astronomers work on it full-time, with the majority of the work being done in the United States. Even if we had more people looking, using better and bigger and more equipment, the smallish rocks a hundred or so yards across are still a threat: it's very difficult to spot them with any reasonable lead time. Many this size are discovered just *after* they pass the Earth, in fact. It's quite possible that the first warning we may get of a small Tunguska-level impact is the flash of light as it streaks across the sky.

So astronomers keep searching, and hoping they'll catch the next impactor with plenty of time to do something about it. The goal is to find 90 percent of all Earth-approaching asteroids bigger than about a thousand yards across by the end of 2008. There are thousands upon thousands of such objects, so the astronomers have plenty of work to do. While the formal 2008 goal was not officially met (it will be eventually), the important thing to note is that, statistically speaking, a large number of asteroids with initially uncertain impact probabilities have been relegated to the "harmless" category.

We've known about asteroids for two hundred years, and it's taken us this long to recognize their danger. The dinosaurs never had a chance.

ARMAGEDDON NOW

Of course, the big difference between us and the dinosaurs is that they didn't have a space program.

You've seen the movie a hundred times: an asteroid miles across is discovered and its orbit puts it on a direct collision course with Earth. If we don't do something, it'll wipe us out. Enter the team of brave hero astronauts/oil riggers/military men. Heroically they launch into space, heroically face down the monstrous rock, and heroically blow it to smithereens, which then rain down harmlessly as gawkers look on.

That sounds, well, *heroic.* There's only one problem: it won't work.

Actually, there are lots of problems with this scenario. For one, there's no guarantee that nuking an asteroid will destroy it. A lot of asteroids are almost solid iron, so throwing a nuclear bomb at one might only warm it up a little.

Even if an asteroid is made of rock, there's no guarantee a nuke will disintegrate it. First, if it's really big, a nuclear weapon may not do all that much damage to it. But it also depends on the asteroid's consistency.

Some asteroids have been found to have very low density, which was initially puzzling. Rock has a density of about two to three grams per cubic centimeter (roughly an ounce per cubic inch, or two to three times the density of water). But some asteroids have lower density than that. An asteroid called 253 Mathilde, for example, which orbits the Sun between Mars and Jupiter, has a density of about 1.3 grams per cc. It must have a texture like Styrofoam. How can that be?

When asteroids were finally observed up close by space probes, they were seen to be heavily cratered. Obviously, asteroids eat their own: they hit each other, leaving giant pits across their surfaces. If an asteroid gets hit hard enough, it'll explode, breaking apart completely. But if it's hit just *softer* than that critical speed, it won't blow apart: the shock from the impact will shatter it in place, like a hammer tapping a crystal egg. The asteroid's own gravity will still hold it together, but it'll be riddled with crevices and cracks. In essence, it's a floating rubble pile.

What would happen if you tried to nuke something like that?

Asteroid expert Dan Durda from the Southwest Research Institute in Boulder, Colorado, wanted to find out. He discovered that the scientific literature on asteroids didn't have much information involving experiments on actual asteroidal material, so he set about to correct that. He obtained some meteorites known to have come from asteroids. One was dense, solid, and rocky, and the other was more porous, more like 253 Mathilde than a chunk of, say, quartz.

He took his sample to NASA's Ames Research Center in California, which boasts the ownership of an unusual gun: it uses compressed air to shoot projectiles at several kilometers per second.

Durda set up his solid specimen in the gun's sights, and slammed a BB into it at five kilometers per second. As expected, the meteorite exploded, disintegrating into hundreds of pieces.

He then put a porous piece of rocky material in the crosshairs. When the projectile hit it, the meteorite *absorbed the projectile and didn't shatter.*

Durda asks, "What if an object like that were coming toward the Earth and you were trying to stop it? . . . How would it respond if we were to throw some sort of a projectile into it at a very high speed to try to break it up? What if you were to try to put a small nuclear device next to it to try to break it up? Would it actually respond in the way you normally think of a solid hunk of rock?

"You put a brick next to you, and take a hammer, and you slam the brick, and it goes flying into pieces . . . that's what you think of when you talk about breaking up an asteroid.

"But you take a sandbag, and you whack it with that same hammer, and nothing happens. It just kind of goes *thud*, and that's the end of it."

That's bad news for us. A rubble pile is pretty good at absorbing damage, and so a nuke won't destroy it. If we see one headed our way, we can bomb the heck out of it, and it'll laugh all the way down to impact.

Actually, many scientists are rethinking the idea of nuking an asteroid. A huge disadvantage of blowing up an incoming threat, even if we could, is that it would create thousands of small potential impactors out of one big one. That may sound better than the alternative (a giant one hitting intact), but even a rock a hundred yards across could easily take out a city. A dozen of those impacting at the same time would be disastrous no matter where they hit. While the explosions would be smaller, they'd be more spread out, scattering damage across the globe instead of confining it to one place.

Durda adds another danger of blowing up small asteroids. "If you take the cosmic composition of a typical asteroid of that size . . . there's enough chlorine and bromine in that object to destroy the ozone layer. So it doesn't matter whether or not that object hits all at once, in one big piece, or if all of that small debris left over from breaking it up into a million pieces still comes raining through and vaporizing into our atmosphere. You're still depositing all of that very harmful stuff into our fragile atmosphere."

It might be better, then, if there's no way to stop it, to just let a smaller asteroid hit us.

That's unsatisfactory, of course, especially if you're sitting in the bull's-eye.

But there may yet be another solution.

One idea is to drop a bomb not *on* the asteroid, but *near* it. Blowing up a bomb next to an asteroid, say, a few hundred yards away, would generate a huge amount of heat and vaporize part of its surface. The solid rock or metal would turn into a gas and expand rapidly, acting like a rocket. It would push a little bit on the asteroid, moving it.

It wouldn't be much, but in space you don't *need* much: every little push adds up. If you blow up several bombs, you can actually generate enough thrust to move the rock significantly. If you move it enough, it'll miss the Earth entirely.

And a big advantage of this technique is that it would work on rubble piles too, though it's not clear exactly how well.

There are some disadvantages to this method. You need a lot of lead time, for one thing. The farther away an asteroid is from impact, the less you have to change its orbit to make it miss us. Most experts think ten years' warning is enough, though they'd be happier with twenty. A century would be just fine. This technique would work best on smaller asteroids since they're easier to move, but a smaller asteroid is also fainter, and thus harder to find. Lead times would be shorter so there would be no room for mistakes. And getting one bomb to an asteroid is hard; getting twenty or more is a lot harder.

Another problem is that it's nearly impossible to know how an explosion would affect the orbit of the asteroid. It might be enough to have the rock miss us, or it might nudge it into an orbit that will hit us on the *next* pass around the Sun.

For example, look at asteroid 99942 Apophis. It's an Earth-crossing chunk of rock about 250 meters across, and is a potential impactor. At that size and mass, it would do considerable damage should it hit, exploding with the force of 900 megatons (more than a dozen times the yield of the largest nuclear weapon ever detonated). Apophis will pass by the Earth on April 13, 2029; there is no risk of impact at that time, but it will pass so close that it will actually be closer to the surface of the Earth than many weather and communication satellites.

The asteroid will approach so close to us, in fact, that its orbit will be seriously affected by Earth's gravity, and just how much its orbit is changed depends on just how close it gets to Earth in 2029. In fact, there is a region of space called the *keyhole* such that if Apophis passes through it, the orbit will be changed precisely enough that on its next return in 2036, Apophis will impact the Earth.

This sweet spot is not terribly big, but our knowledge of Apophis's exact trajectory isn't good enough to completely preclude the asteroid's passing through it. The odds are incredibly low, maybe less than 1 in 45,000, but it's worth investigating.

And what if it turns out that Apophis will glide right through the keyhole? We'll have just seven years to move it enough to miss us. A better idea is to prevent it from passing through the keyhole in the first place. If we get to Apophis before 2029, then we hardly have to nudge it at all; calculations show that changing its velocity by even a few thousandths of an inch per hour would work. So you might think that a well-placed nuclear weapon would do the trick.

Unfortunately, it won't. That keyhole isn't alone: there are dozens of keyholes, thousands. That first keyhole is just for a return of Apophis in seven years, but other keyholes will bring it back in ten years, twelve, twenty . . . instead of saving us, a detonation just buys a little bit of time, and there's no guarantee that we can move it away from some other keyhole—or knock a chunk or ten of it into another keyhole.

Controlling the resulting orbit is a key issue, and blowing up a nuclear weapon is not exactly subtle.* We need more fine-tuning on asteroid steering.

RAMMING SPEED

By now it may have occurred to you that maybe we don't need a bomb. The impact of an asteroid on the Earth releases energy like a bomb, so why not try impacting the asteroid itself? If we hit it hard enough with some sort of impactor, we won't need a nuke.

---

* It's also illegal. A 1963 test ban treaty (see chapter 4) forbids the detonation of any nuclear weapons in space. However, one would assume that international treaties might be set aside temporarily given the total annihilation of the human species as the alternative.

There is a very big advantage of this method: we've done it before. On, appropriately enough, July 4, 2005, NASA's Deep Impact probe slammed into the comet Tempel 1, creating a flash seen by hundreds of scientific instruments across the world. The impactor was an 800-pound block of copper, which was steered into the comet at over six miles per second. The resulting explosion was the equivalent of about five tons of TNT detonating. The size of the resulting crater is unknown; the flash and debris hid the impact from the spacecraft's camera.

Steering a probe into an object moving at several miles per second is an engineering triumph. There were no second chances, and even the exact shape of the comet nucleus was unknown until the probe got there.

On the other hand, the comet itself was three by five miles in size, which is pretty big. Had it been a small asteroid, it's unclear whether the NASA engineers would have been able to hit it. Still, it was a first shot, and a successful one. Much was learned from the attempt that can be applied to ramming a potentially dangerous asteroid.

But it must be stressed that the impacting scenario suffers from most of the same issues as bombing an asteroid: it might shatter the asteroid, producing many smaller impactors; if the asteroid is porous it will simply absorb the impactor; and again we cannot control the resulting orbit, so we might just be pushing it into some other future impact event. While this might change the orbit enough to miss the Earth, it can't be known in advance just how much, and in this game inches matter.

VIRTUAL TETHER

Still, there may be other ways to rid ourselves of a potential planet-buster. Perhaps, instead of blowing one up, we can instead gently *persuade* the asteroid to change its trajectory.

The B612 Foundation—named after the asteroid home of Antoine de Saint-Exupéry's titular Little Prince—is, for lack of a better description, a kind of doomsday think tank, consisting of dozens of scientists, engineers, and astronauts whose express purpose is to figure out a way to save humanity from the threat of giant impacts. The foundation has held meetings, written papers, and had members (such as Apollo 9 astronaut Rusty Schweickart) testify before Congress about doomsday rocks.

Their Web site reads like a science-fiction novel, full of ways to stop an asteroid from hitting us. However, the emphasis is on the science. While many of the methods would be difficult to execute and are clearly only in the very early stages, others involve mature technology or adapting what we already have.

For example, one method is to physically land a rocket on an asteroid, secure it in place upside down, and then start firing it. Over time, the thrust will push the asteroid into a new orbit, making the rock miss us.

This may be the safest method, and it certainly makes sense, but in reality it would be pretty hard to do. For one thing, it's not entirely clear how you would secure the rocket to the surface of the asteroid. What if the surface is powdery, or it's a rubble pile, or it's metal? For another, every asteroid spins, which means you can only fire the rocket for short periods of time when it's pointed in the right direction. That means you need more lead time, and in many cases time is precious. Worse, some asteroids tumble chaotically, and for those a rocket would be nearly useless.

These problems kept the B612 Foundation members thinking . . . and they came up with an answer that is really quite surprising. What if you don't land the rocket at all?

Asteroids are small compared to planets, but they still have mass. And any object with mass, said Isaac Newton, has gravity. The rocket itself has mass, and therefore gravity as well. So imagine this: a rocket is placed in a parking orbit *near* the asteroid, but not physically *in contact* with it. The asteroid's gravity will pull on the rocket, making it fall toward the asteroid. In the same way, the rocket's mass will pull on the asteroid. Now the rocket is fired, but very, very gently, just

An innovative idea in asteroid impact mitigation is to use the gravity of a small spacecraft to move a hazardous asteroid out of harm's way. Given enough lead time, this is a very precise method of altering an asteroid's orbit.

DAN DURDA (FIAAA) AND THE B612 FOUNDATION

enough to counteract the fall toward the surface. The result is an asteroid tug. But unlike a barge or a tugboat on Earth that uses ropes to haul other boats, the rocket is virtually connected to the asteroid by gravity. Over time, the gravity from the rocket pulls the asteroid into a safe orbit.

But of course there are technical difficulties with this method too; there always are. The rocket cannot fire straight down, toward the asteroid, because it will push the asteroid back, negating the effect of the tug. So the rocket will have to be tilted outward, firing at an angle away from the asteroid. That means that pairs of rockets are needed to balance each other and keep the tug from spinning out of control.

The amazing thing about this method is that in some cases, the mass of the tug need not be all that much. For 99942 Apophis, for example, a tug massing only *a single ton* can be effective in moving the asteroid away from the keyhole even if it gets to the rock only two years in advance. To be fair, in general it will take longer to move an asteroid away from an impact trajectory; for Apophis we need only move it so it misses a small region of space, but for a direct-impact trajectory the asteroid needs to miss a whole planet. That means moving the orbit thousands of miles, which in turn means a longer lead time (or a more massive tug). One current thought, developed by Schweickart, is a hybrid solution: using a kinetic impactor (literally whacking it with another rock) or nuke to move the asteroid out of the immediate threat zone, then using the gravity tug to fine-tune the orbit so that we don't get a surprise a few orbits later.

Still, as promising as these technologies sound, we need to be honest with ourselves. We currently don't have the technology to implement *any* of these methods. We're close—maybe only a few years from developing the gravity tug—but even lobbing a nuke at an asteroid is pretty difficult. A report written in 2007 from NASA to Congress suggests that sending an impactor to an asteroid is currently our only workable option.

But that is due to current knowledge and current technology. The B612 Foundation is hoping to incrementally test technology that will prevent an asteroid from hitting us. Even better, some of their ideas, like the gravity tug, allow us to manipulate the orbit of an asteroid any way we want. We might even be able to nudge one into a safe orbit around the Earth. It would be far too small for its gravitational pull to affect us, but close enough that we might be able to set up mining operations on it. That might sound far-fetched, but some estimates show that the metals in even a small asteroid could be worth trillions of dollars. That would make a mighty tempting target for industry.

Instead of its aiming at us, we would be well suited to aim for *it*.

COMET WHAT MAY

There is still another lurking problem about which we need to be aware. Asteroids tend to have nice, predictable orbits. They are dead hunks of

In this artist's work, Space Shuttle astronauts see a comet nucleus dozens of miles across impact the Earth. Orbiting astronauts may be the only survivors; the Earth they would eventually return to would be devastated by such an event.

DANA BERRY, SKYWORKS DIGITAL INC.

rock and/or metal, so once we observe them for a while, we can predict their orbits for decades.

But asteroids aren't the only threat. Comets are lovely, wondrous specters in the sky. Unlike asteroids, comets are like dirty snowballs: rock, gravel, and dust mixed in with ice holding it all together. When they get near the Sun, the ice melts.* Many comets have pockets of

* Technically, it never becomes a liquid; it goes directly from a solid to a gas in a process called *sublimation.*

ice under the surface, and when those sublimate the gas vents out in a jet. This acts like a rocket, pushing the comet around. If the comet is spinning—and most are—this means the comet will get pushed around randomly. That makes it extremely hard to accurately predict their orbits, and that much harder to land a rocket in them, or to use a gravity tug.

And it gets worse. The solar system looks something like a DVD seen edge-on: the planets orbit the Sun in the same plane. Asteroids too tend to stick to that plane. That means looking for them is a lot easier; we only need to keep checking the same parts of the sky.

But comets are wild cards. They aren't confined to the solar system plane, and can come literally from any part of the sky. This can significantly cut into the lead time we have to do something about a killer comet approaching Earth. While we might have decades of notice for an asteroid impact, we might only have a few years for a comet. Even comet Hale-Bopp, which was one of the brightest ever seen, and which delighted hundreds of millions of people, was only discovered about two years in advance of its passage of Earth. Had it been aimed at us, there wouldn't have been a damn thing we could have done about it. Hale-Bopp's nucleus—the solid part of the comet—was twenty-five miles across. Had it hit, it would have made the asteroid impact that wiped out the dinosaurs look like a wet firecracker.

But even a small comet could have a disastrous, well, *impact.* Assuming it wasn't confused for a sneak attack of some kind, the direct consequence of a small impact or Tunguska-like airburst over a city could lead to thousands of deaths and billions of dollars of damage. If it happened over a major city or economic landmark—New York City, California's Central Valley (where much of the nation's fruits and vegetables are grown), Tokyo—the results could be far worse. The good news is that long-period comets like Hale-Bopp represent less than a few percent of the overall impact hazard, and most short-period comets are easy to spot.

ODDS AND ENDS

So how big a danger are asteroid and comet impacts?

Statistically speaking, you're not going to like the answer: the odds of getting hit are 100 percent. Yes, really. Given enough time, and if we do nothing about it, there *will* be impacts, and one *will* be big.

But the key part of that sentence is the "if we do nothing" part. The point is, we *can* do something. While the techniques described here sound like something out of a movie, they are all possible. Technically they'll be tough, and they'll be expensive. But the stakes are pretty high: global survival versus utter annihilation.

I think that given this, it's about time we took these science-fiction ideas and made them science fact.

CHAPTER 2

Sunburn

IT'S JANUARY, THE DEAD OF WINTER ON THE NORTHERN *hemisphere of Earth. During the short days, the Sun makes a desultory appearance low in the sky, only to sink below the horizon again a few short hours later. It can barely warm the planet, it seems. With the chill in the air, people don't give the Sun a second thought. They wouldn't even think it had much of an impact on their lives.*

They're about to be proven quite wrong.

The Sun is nursing a cosmic hangover. It has undergone some violent paroxysms over the past few years, erupting multiple times, sending tremendous blasts of matter and energy into space. Through sheer chance, these had mostly missed the Earth. The worst thing that had happened was one eruption nicking the Earth, causing beautiful aurorae at both poles, and disrupting some radio communications: an annoyance, but easily offset by the stunning display of northern and southern lights.

Things are on the decline now, and the Sun appears to be calming down. Scientists are just starting to think they can breathe easier.

They're therefore caught by surprise when a vast group of sunspots peeks over the edge of the Sun. Sunspots are dark blotches of cooler material, caused by kinks and twists in the Sun's magnetic field, and they are harbingers of solar activity. Scientists scramble to observe the sunspot group, bringing a fleet of ground-based and orbiting telescopes to bear on the star. They are greeted by an ugly sight: the Sun's surface is gnarled, twisted, blackened, defaced by the spots. This group is a whopper, as big or bigger than the largest groups seen in 2003, which scientists still buzzed about.

For over a week astronomers nervously watch the active region, measuring its size, shape, and magnetic activity. The latter appears to have settled down, which could indicate either that the magnetic field is fading or that it is building up like a volcano.

They soon get their answer. The sunspots, normally dark, brighten tremendously in seconds, and stay bright for many minutes. At the same time, orbiting solar telescopes note wild magnetic fluctuations on the Sun, and minutes later are flooded with high-energy X- and gamma rays. Astronomers on the ground monitoring the orbiting observatories see unprecedented energy blasts, with measurements off the scale, when, suddenly, the data flow stops. Bewildered for a moment, they check their equipment, but then realize the problem is not on the ground, but in the sky: the huge influx of energy has fried their astronomical satellites.

Knowing that commercial satellites are at grave risk as well, the scientists make frantic calls to other observatories, but find the phones aren't working either. Turning to their computers, they try e-mail, instant messaging, voice-over-Internet, anything, but communication is impossible. Nothing is working. Then their power goes out, and they realize things are about to get much worse.

Shortly after the flare, the Sun unleashes another blast, this time in the form of a brutal wave of subatomic particles. Traveling at phenomenal speed, the wave reaches the Earth, where it slams into and flows over the planet's protective magnetic field. Submerged in the electromagnetic mayhem, satellite after satellite dies from an overdose of sunburn.

The effect reaches the ground as well. Transmission wires are suddenly overloaded with current, heating up, sagging, and snapping. Transformers are overwhelmed, exploding. Workers at electrical stations across the United States and Canada are snapped out of their routine and suddenly find themselves struggling valiantly and frantically to keep up with the cascading disaster, but it's hopeless. Station after station goes down. Power goes out first to the U.S. Northeast, but within seconds the grid goes down in an expanding wave. Quebec, Boston, New York City, Philadelphia . . . minutes later, a hundred million people are without power at night in the dead of winter. They wake up the next morning to icy homes, without electricity, and with no means of finding out what happened.

Within hours, over half the planet is without power during one of the coldest winters in recent memory. Thousands die the first night, and many more follow over the next few weeks. The military jumps in, doing what it can to help those in need, but the sweep of the disaster is simply too broad. The number of deaths is staggering, an epic catastrophe on a scale unseen for a century. The economic impact alone is measured in the trillions of dollars, and entire nations go bankrupt.

Eventually, the Sun calms down. The active group of sunspots fades away. But magnetism on the Sun is fiercely complex. Within a few weeks, tangles and interconnections reappear in the solar magnetic field. Just as things on Earth start to settle, and people are able to bury the dead, another group of ugly sunspots begins to build on the star's surface.

MY SUN, THE STAR

An occupational hazard of being an astronomer is getting free astronomy textbooks in the mail. Like e-mail spam (but tipping the scale at ten pounds), they come unannounced, and generally wind up in a used bookstore collecting dust (the real-world equivalent of the spam filter).

I can't resist thumbing through them. I torture myself this way, knowing that I'll find some odd chapter arrangement, some scientific error, some small turn of phrase that will irk me in some way. And always, without fail, I find it in the section about the Sun. Invariably, there will be some permutation of this sentence: "The Sun is an ordinary, average star."

If you decide to read only this chapter and then close this book forever, then please walk away with just one thing: *the Sun is a star,* with all that this implies. The Sun is a mighty, vast, furiously seething cauldron of mass and energy. The fires in its core dwarf into microscopic insignificance all the nuclear weapons ever built by mankind. A million Earths would be needed to fill its volume, and the light it emits can be seen for trillions upon trillions of miles. Invisible forces writhe and wrestle for control on its surface, and when it loses its temper, the consequences can be dire and even lethal.

That is what it means to be an "ordinary" star.

Let's be clear—there *are* lots of stars like the Sun, and if you phrase it carefully, then sure, the Sun is average. The smallest stars have roughly one-tenth its mass, and the largest have a hundred times its mass, so the Sun is somewhere near the low end of the range. But this neglects the actual *population* of stars: low-mass stars are far, far more common than their hefty brethren. More than 80 percent of the stars in our galaxy are lower-mass than the Sun. Roughly 10 percent have the same mass as the Sun, and 10 percent have more. So really, in a standardized cosmic test, the Sun scores pretty well. Maybe a B+.

Of course astronomers—and I count myself guilty here as well—do love to use diminutive adjectives when describing low-mass stars: *dinky, tiny, feeble.* But that's hardly fair, either: even the smallest star is far, far larger than Jupiter, and Jupiter is pretty big; three hundred Earths would fit inside it, so even a small star is a huge object.

And yet the Sun is larger in size than the majority of stars in the galaxy: their median diameter is about a tenth that of the Sun. So even on a cosmic scale the Sun is *big.*

On a *human* scale, as you can imagine, it's a scary, scary place.

The Sun is about 93 million miles away. If you could build a highway and drive there, it would take over 170 years. Even an airplane would take two decades to fly to the Sun if it could.

And yet . . . imagine it's summer and you're standing outside. You turn your face up to the Sun. Feel the warmth? Sure! The Sun is so bright you can't even look at it. And if you stand there for more than a few minutes you risk damaging your skin.

The Sun's fearsome power is generated deep in its core, where a controlled nuclear reaction is taking place: the Sun is continuously fusing nuclei of hydrogen together to create helium nuclei. Every time this reaction occurs a little bit of energy is given off, and in the Sun's core the reaction happens a *lot:* every second of every day, the Sun converts *700 million tons* of hydrogen into 695 million tons of helium.

The missing 5 million tons get converted into energy, via Einstein's famous equation $E=mc^2$, which shows that mass and energy can be converted back and forth into one another, and that a tiny bit of matter produces a whopping amount of energy. Five million tons is a huge amount of matter, the equivalent weight of seven fully loaded oil supertankers . . . and the Sun chews through that much hydrogen *every second.**

The energy created every second in the core of the Sun—equal to the energy it emits from its surface—is the equivalent to the detonation of *100 billion one-megaton nuclear bombs.* This is 200 *million* times the total explosive yield of every nuclear weapon ever detonated on, below, and above the surface of the Earth . . . and the Sun does this every second of every day, and will continue to do so for billions of years yet to come.

Some people like to say the Sun is essentially a giant nuclear bomb, but that's misleading: a bomb explodes.† But the Sun doesn't explode, because it has a lot of mass. This means it has a lot of gravity, which balances the energy it generates. The heat produced makes the Sun want to expand (like a hot-air balloon expands), but the Sun's own gravity holds it together. It's a balancing act; in fact, a good definition of a star is a ball of gas with nuclear fusion in its center held together by its own gravity.

---

* Not to worry, the Sun won't run out of hydrogen anytime soon—5 million tons sounds like a lot, but it's only 0.00000000000000000025 percent of the Sun's mass. We have billions of years of fusion still ahead of us.

† You might say that's a bomb's distinguishing feature.

But just because the entire Sun doesn't explode like a bomb doesn't mean that explosions don't happen. In fact, the Sun is capable of epic eruptions; but they're not nuclear in nature, they're magnetic.

CURRENT EVENTS

When I was a kid (and sure, I'll admit it: even today), I was fascinated by magnets. I had a few different kinds, and I would play with them constantly. I read a lot about magnetism, and in one of my books it said magnetism could be destroyed by heat. I (carefully!) held a bar magnet in a candle flame for a few minutes, and sure enough, after that it wouldn't attract nails or needles anymore.

I was also something of an astronomy geek even then, and I had a book that talked about the magnetic field of the Sun. I remember being confused by this: how could the Sun have a magnetic field if it was so hot?

What I didn't understand is that there is more than one way to create a magnetic field. Simply put, a magnetic field can be generated by moving electrical charges. When you turn on a light, for example, electrons (subatomic particles with a negative charge) flow through a wire from the wall to the light. This motion produces a local (temporary) magnetic field around the wire. When you turn off the light, though, the flow of electrons stops, and the magnetic field collapses.*

This has a very interesting—and useful—effect. If an electrically conductive object like a wire moves through a magnetic field, an electric current will flow along the wire. This current, in turn, generates its own magnetic field. If the current moves in just the right way, its magnetic field will reinforce the magnetic field already there and you get a self-sustaining system.

---

* You can check this yourself with a simple compass. Find a house lamp or some other appliance connected to a wall outlet. Put the compass near the wire, and turn the appliance on and off. The needle will move, influenced by the temporary magnetic field.

However, this only works if there is an outside source of energy to make things move. For example, you could use a crank to make a coil of copper wires rotate inside a magnetic field (generated by a permanent magnet). Your arm supplies the outside energy. Or, if you're smart, and you want to make a lot of electricity, you stick this getup near a source of flowing water—say, inside a dam—and make giant turbines composed of copper that spin as water flows past them . . . which is precisely how hydroelectric power plants work. A system that converts mechanical energy to electromagnetism in this way is called a *dynamo.*

The Sun is just such a dynamo. Its interior is hot: so hot, in fact, that electrons are stripped off their atoms, allowing them to flow more or less freely. An atom that is missing one or more electrons is said to be *ionized.* As these electrons move in the ionized gas, they generate magnetic fields.

If the Sun were just sitting there in space, a nonmoving and nonrotating ball of hot gas, the electrons inside would move around higgledy-piggledy, and all those individual magnetic fields generated would be oriented in random directions and cancel each other out. But the motions of the electrons in the Sun are far from random. For one thing, the Sun spins on its axis once a month, and that can create streams of gas in its interior. This preferred direction of motion for the electrons means that their individual magnetic fields can build on one another like creeks all flowing into a river, creating a larger magnetic field.

If it were just that simple, scientists would understand everything about how the Sun works. But in reality the Sun is incredibly complicated, with a vast system of moving gas inside it. The heat from the core makes gas above it rise,* generating towering conveyor belts of gas over 100,000 miles high, moving up and down inside the Sun. Other rivers of gas move around it like the jet stream does on Earth, and yet another

---

* This process, called *convection,* is what causes hot air to rise and cool air to fall, and also can be seen when you heat a pot of water on a stove.

set of streams flows north and south as well. When taken all together, the Sun more closely resembles a ball of writhing worms than a simple sphere of gas. It's like a street map of Tokyo, but in three dimensions and changing with time as well. Because of this, the magnetic field of the Sun is a nightmare as well, making it ferociously difficult to understand. On the positive side, though, it also keeps a lot of solar physicists off the streets.

All of this together is what creates the Sun's dynamo. The charged particles inside the Sun are moving in currents. These currents move inside a magnetic field, so the currents themselves generate a magnetic field, and the whole thing is self-reinforcing. The crank, in this case, is the Sun itself, with its own rotation providing the mechanical input energy needed to generate the dynamo. The Sun is huge and massive, so there is a vast amount of rotational energy to tap into. The solar magnetic field is created at the cost of the Sun's spin, but it will take billions of years for the energy loss to result in a noticeably slower spin.

The Sun's magnetic field is complicated and interesting, and by interesting, of course, I mean dangerous.

Or had you forgotten the title of this book?

MAGNETIC BUBBLE, COIL AND TROUBLE

Earlier, I mentioned that a star can be defined as an object with fusion in its center, whose tendency to expand due to energy production is balanced by its gravity.

Stars are a study in balance in this way. If gravity were weaker, they'd expand or explode. If their energy generation were a little weaker, they'd shrink or collapse (more about both of these in later chapters). Their rotation, chemical composition, gravity, heat, pressure, and yes, magnetic field all combine in exquisite balance to produce a stable star.

But sometimes things get out of whack.

When a simple magnetic field is illustrated, you usually see a set of lines emerging from the poles of the magnet, connecting one pole to the other. The field lines of a bar magnet, for example, look something like a doughnut. These magnetic field lines are useful to visualize the strength of a magnet: where the lines are bunched up together (like near the poles of a bar magnet), the magnetic field is stronger; where they are spaced out the field is weaker. If you place a small bar magnet inside the magnetic field of a larger magnet, the smaller one will align itself along the larger's field lines. That's why a compass points north; the needle is a magnet, and it aligns itself along the Earth's magnetic field lines.

Things get complicated if the magnet is not a simple shape. If you bend a bar magnet, the field lines will bend as well. If you take a dozen magnets, a hundred, and throw them together, the field lines can get very distorted, because each bit of the magnetic field is attached to the object generating it. Mess with one and you affect the other.

The magnetic field of the Sun is generated by moving currents of gas—currents that get twisted, distorted, and bent around just like rivers on the Earth. These field lines may be generated beneath the surface of the Sun, but they don't stay down there; they pierce *through* the surface, looping upward and back down into the interior in an incredibly complex, interwoven, and interconnected way. These magnetic field lines can really get their knickers in a twist, becoming entwined and entangled. When this happens, there are profound changes on the surface of the Sun.

For one thing, since the field lines and the gas are coupled, when the lines get tangled and compressed, the gas has a harder time moving around as well. It's like a giant net is thrown over the gas, preventing it from moving freely. Hotter gas welling up from below can't reach the surface, and regions where the lines are particularly dense begin to cool off. Since the brightness of the Sun is due to its temperature, a cooler region becomes dimmer, forming a dark area on the Sun called a *sunspot.* Because sunspots are inherently magnetic phenomena (they are really a cross section of the magnetic field lines where they intersect the surface of the Sun), they always come in pairs with reversed magnetic polarity: one is like a magnet's north pole, and the other is the south pole.

Sunspots can be small, barely visible to telescopes on Earth, and they can be huge, dwarfing the Earth itself, with some so large that they can be seen by the naked eye when the Sun is on the horizon.*

In fact, it was the observation of sunspots that first keyed astronomers into the Sun's magnetic field. Heinrich Schwabe was a solar observer in the early nineteenth century who counted the number of sunspots every day for decades. He discovered that the number of spots waxes and wanes with a period of about eleven years from peak to peak—we now call this the *sunspot cycle.* At the time of the maximum, there can be well over a hundred sunspots on the Sun, but at the minimum that number drops to essentially zero.

Schwabe decided to publish his results in 1859, and it was quickly determined that the times of peak sunspot number also corresponded to the times of peak magnetic activity on the Earth, indicating a connection between sunspots and magnetism. In 1908, the astronomer George Ellery Hale discovered that the magnetic fields in sunspots can be thousands of times stronger than the Earth's, indicating the presence of intense energies being stored there.

---

* When the Sun is on the horizon, its light passes through more air than when it is overhead, dimming the sunlight considerably and making it easier to see sunspots. At this point, you might be expecting me to exhort you to never ever look at the Sun. However, surprisingly, there has never been a reported case of permanent total blindness caused by looking at the Sun. It *is* possible to damage your eyes looking at the Sun—for example, using cheap sunglasses that dim visible light but not ultraviolet, or looking at the Sun when your pupils have been artificially dilated with drugs—but it's actually rather difficult to do, and in general the eye heals quickly. I don't recommend it since damage *is* possible, but it's unlikely and certainly not worth the hysterics it garners. Having said that, I will point out that looking at the Sun through binoculars or a telescope is in fact *incredibly* dangerous, since they concentrate sunlight. The only 100 percent safe way of looking at the Sun with optical aids without risking boiling the fluids in your eye is to project its image on a piece of paper. There are other, more expensive methods, but this one is easiest. And nothing is more expensive than losing an eye.

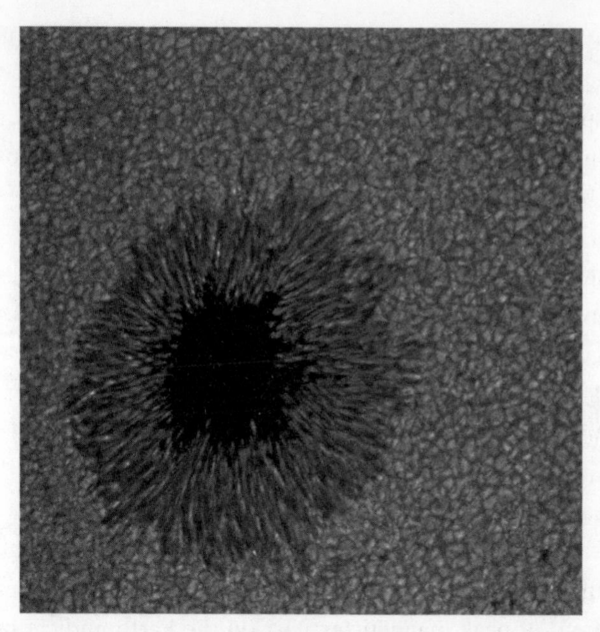

This is a typical sunspot, appearing darker than the surrounding solar surface because of its cooler material. This particular spot is far larger than the Earth. The graininess of the Sun's surface around the spot is caused by convection, rising currents of hot material that cool and sink back down into the Sun.

STANFORD-LOCKHEED INSTITUTE FOR SPACE RESEARCH AND BIG BEAR OBSERVATORY

Which brings us back to balance. As the magnetic field lines tangle up, there is a balance struck between the pressure built up by the magnetic energy stored in them and the tension that exists in the lines. Imagine the magnetic field lines are like steel coil springs, all tangled together and interconnected. The springs are compressed and want to expand, but the tension of the intertwined mess keeps them from springing back. Now keep compressing them and adding more springs, again and again. The energy stored up would get pretty impressive.

What happens if you take a bolt cutter and snip one of the springs? Right. Better stand back.

The same thing happens in a sunspot—in fact, much of the physics is pretty similar to a convoluted mess of coiled springs, with the analogous tension and pressure. As the field lines get more entangled, and

Loops of extremely hot material flow up from the Sun's surface, following along the magnetic field lines. When the loops get tangled or twisted, a flare or coronal mass ejection can be triggered.

TRACE TEAM/NASA

more are added, the pressure builds up. Sometimes the pressure is relieved early in the process, and not much happens. But other times it builds, and builds . . .

Something's gotta give.

Eventually, something does. The field lines emerge from the Sun in tall, graceful loops, with one footprint being the magnetic north pole and the other the south. If the gas flow zigs instead of zags, for example, the footprints can be brought together, or twisted past each other. The pressure in the coil goes up, but the tension can't compensate. The line snaps.

There is a lot of energy stored in the field line (just like the energy stored in a spring). When it snaps—what solar physicists call *magnetic reconnection*—the energy is released. A *huge* amount of energy. The explosion is titanic, but in general constrained to a local region, causing what's known as a *solar flare.*

A FLARE FOR DANGER

By coincidence, a solar flare was first observed in 1859—the same year Heinrich Schwabe published his discovery of the sunspot cycle.

On September 1, 1859, astronomers Richard Carrington and Richard Hodgson were independently observing the Sun. Before their eyes, a small part of its normally calm disk suddenly exploded in intensity, becoming far brighter. This burst of emission lasted for five minutes,

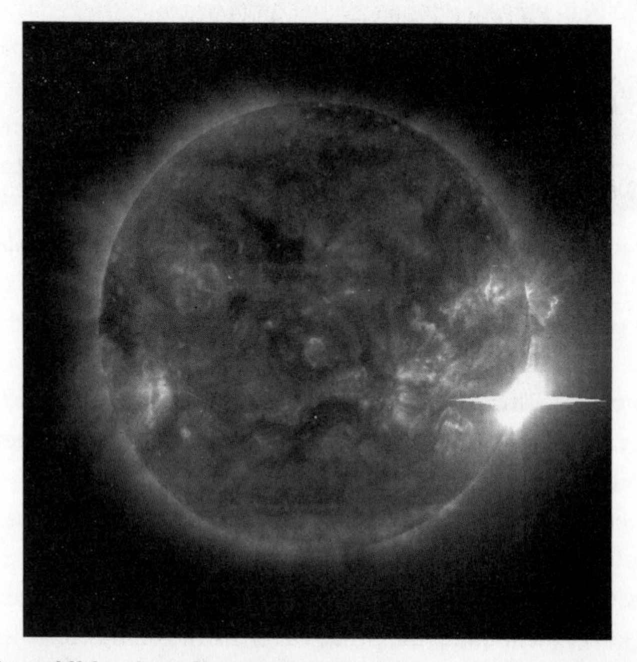

The Solar and Heliospheric Observatory detected this massive flare from the Sun on November 4, 2003. It was one of three huge flares that surprised scientists that day; no such string of events had ever been witnessed before. They marked one of the most active weeks for the Sun ever recorded.

SOHO (ESA & NASA)

and even to this day may have been the most luminous flare ever observed. Within a few hours of the observations of the flare, magnetometers (instruments that measure the strength and direction of a magnetic field) on Earth went crazy, registering huge fluctuations in the Earth's magnetic field.

They didn't know it then, but at that moment the study of space weather was born.

They also couldn't have known that the flare was caused when tangled magnetic field lines on the surface of the Sun suddenly realigned themselves. The energy stored in them was released like a bomb—the equivalent of *15 billion one-megaton nuclear weapons,* or 10 percent of the total energy output of the Sun every second concentrated into one spot—hurling high-energy photons (particles of light) and subatomic particles both upward into space and downward onto the surface of the Sun. A typical flare from the Sun ejects billions of tons of subatomic particles outward at speeds that reach five million miles per hour—and in 2005, one extraordinary flare launched a blast of protons that reached the Earth in just fifteen minutes, indicating they were traveling at *one-third the speed of light.* These subatomic particles blast outward, generally straight out from the center of the flare. Because of this, the particles launched upward and outward from the flare are generally not a problem to us on Earth: they are focused enough that they usually miss us, causing no grief.

But along with the particles shot into space, a huge pulse of particles is shot down, onto the surface of the Sun. This heats the gas there tremendously, and creates an incredibly strong pulse of light. Now, that may not sound like a big problem; after all, how bad can light be?

Bad. But it depends on the kind of light.

What we call "visible light" is a narrow slice of a much wider range of electromagnetic radiation. Infrared light, for example, has less energy than visible light, and radio waves have less energy still. Ultraviolet (UV) light has *more* energy than the light we can see. Still higher-energy light is X-rays, and on up to gamma rays. UV, X-, and gamma rays are dangerous in large quantities. Each photon carries so much energy that it can radically alter any atom it hits, stripping off the atom's electron, ionizing it.

Flares give off a *lot* of this kind of light. And unlike the particles of matter emitted in a solar flare, this light spreads out. A flare on the edge of the Sun's disk will almost certainly miss us with its particles, but *any* flare *anywhere* on the visible surface of the Sun is a potential danger because of the high-energy light it emits.

Picture a solar flare on the Sun: the tangled magnetic field lines over a sunspot suddenly snap, rearranging themselves, and releasing their energy. They heat the local gas up to millions of degrees, and a blast of X-rays surges outward.

Traveling at the speed of light, the high-energy radiation takes a little over eight minutes to travel the 90 million or so miles to the Earth. When it does, it slams into everything in its way: satellites, astronauts, and even the Earth's atmosphere.

On the Earth's surface, we're protected from this onslaught by the thick air over our heads. But an astronaut in orbit is essentially naked, exposed to the wave of photons. A spacewalker caught by surprise will absorb many of the incoming X-rays, getting the equivalent of hundreds or even thousands of chest X-rays in a single flash.

X-rays are dangerous because when absorbed, they deposit all their energy into tissue. This can lead to cell and DNA damage. When DNA is damaged, mutations can occur that can (but do not always) lead to cancer.

Radiation absorption is measured in units called *rems.** Natural radiation coming up from the Earth's surface surrounds us all the time; you get a dose of about 0.3 rem per year just by existing on the Earth. In high-altitude locations, like Denver, that can be as high as 0.5 rem due to both terrestrial and extraterrestrial sources. A dental X-ray, by comparison, gives a dose of about 0.04 rem, one-tenth of your normal

---

* This stands for *roentgen equivalent man* where a roentgen is a measure of an amount of radiation. A source may give off a certain number of roentgens of radiation, but the amount that gets absorbed by human tissue is measured in rems.

annual background dose. The U.S. government has guidelines for employees who work in elevated radiation environments: the maximum safe whole-body dose is set at 5 rems per year.

A mild flare may expose an astronaut to several dozen rems of radiation. While that sounds bad, in fact the body can heal itself fairly well after such a one-time radiation dose. Cells heal, and small amounts of damaged DNA can be eradicated by the body's natural defenses. That's not to say it's fun: the problems associated with this kind of dose are irritated skin and a higher risk of developing skin cancer or other forms of cancer. Male astronauts might also experience a temporary sterility lasting for a few months, and hair loss in both sexes is possible.

But if too much tissue is damaged, the body cannot heal itself. In a major flare, an astronaut could absorb hundreds of rems of X-rays. This can be fatal: there is simply too much cell damage for the body to repair itself. Over the course of several hours and days the astronaut suffers a slow death as cells die, the intestinal lining sloughs off, ruptured cells leak fluid into their tissue . . . the effects are horrifying. NASA takes this threat *very* seriously. When a flare is seen on the Sun, astronauts on the International Space Station retreat to a section that is more protected, letting the station itself absorb the radiation to safeguard the humans inside.

When astronauts return to the Moon they'll have to deal with this as well. Lunar rock is an excellent absorber of radiation, so it's likely that lunar colonists will cover their habitats with two or three yards of rock and rubble. It's not as romantic as glass domes on the surface, but being able to actually survive a flare may take precedence over our preconceived notions of what a colony should look like from watching science-fiction movies.*

In a major flare, though, not just humans are in danger: our satel-

---

* Going to Mars is even more difficult, since it takes months to travel there. Radiation from flares will be an even bigger priority. Because of its mass, rock makes an inconvenient shield for interplanetary travelers. NASA and other space agencies are busy trying to solve this problem so that trips to Mars can become a reality.

lites can be fried as well. When an X-ray or a gamma ray from a flare hits the metal in a satellite, the metal becomes ionized. A very high-energy gamma ray can ionize many atoms in the satellite, causing a cascade of electron "shrapnel" to fly off the atoms. Remember, moving electric charges create a magnetic field. This sudden strong pulse of magnetic energy can damage electronic components inside a satellite (just as a magnet can damage your computer's drive). The electrons themselves might short-circuit the hardware too.

Many civilian satellites have been lost in solar flare events. Military satellites are in many cases protected from this damage, and such radiation-hardened satellites can still operate even if there is a major flare. The effects of a nearby nuclear blast are similar to those of a flare, so these satellites may also survive a nuclear detonation in space (as long as debris and heat from the blast doesn't get them).

Moreover, the Earth's atmosphere absorbs the incoming high-energy light. While that protects us on the surface, the upper atmosphere can heat up from this and "puff up" like a hot-air balloon. If the atmosphere expands enough, it can actually reach the height of some satellite orbits. A satellite normally orbiting in a near-vacuum environment may suddenly find itself experiencing drag as it plows through the very thin extended atmosphere. This lowers the satellite's orbit, dropping it into even thicker air, where it drops more, and so on. Even if it survives the initial flare, it may still be destroyed when it burns up in the Earth's atmosphere! Many low-orbiting satellites are lost every solar cycle because of this effect. The American space station Skylab was destroyed this way in 1979.

Because of this, space agencies and commercial satellite owners watch for flares very closely. Flares are linked to the eleven-year sunspot cycle, tending to occur on or around the solar sunspot maximum, though for reasons still not well understood, the most energetic flares usually happen about a year after maximum. Incidentally, the 1859 flare, perhaps the brightest of all time, occurred a year or so *before* the sunspot maximum.

That flare induced quite a bit of magnetic activity on the Earth.

While the flare itself probably did have some direct effect on the Earth, it's now thought that it had some help.

HALO, HOW YOU DOING?

Normally, there is a relatively constant flow of material from the Sun. Called the *solar wind,* it's a stream of subatomic particles accelerated by the usual suspect: the solar magnetic field. The solar wind blows off the Sun in all directions, and continues outward for billions of miles, well past the orbit of the Earth around the Sun. Near the surface of the Sun, the particles can be seen as a faint pearly glow called the *corona.* The corona is incredibly hot—billions of degrees—but extremely tenuous, like a laboratory-grade vacuum. But over the trillions of cubic miles of solar surface, even something so diffuse can add up to a lot of mass. Astronomers think of the corona as the atmosphere of the Sun, so, in a very real sense, we live in the atmosphere of a star.

This has some disadvantages. Atmospheres sometimes have bad weather.

When a flare erupts from the surface of the Sun, needless to say, it tends to have an effect on its environment. The blast of energy and particles from the flare goes upward, of course, away from the Sun, but it also goes *downward,* onto the surface. This creates a seismic wave on the surface of the Sun with tens of thousands of times the energy of the strongest terrestrial earthquakes. The Sun's surface ripples as waves of energy are slammed into it. The magnetic field lines surrounding the energy get an enormous jolt as well, and many times this is enough to disrupt them. The lines going in and out of the Sun's surface in the area reconnect, release energy, and disrupt more lines around them. More and more energy is released as the effect spreads and more lines reconnect.

As this occurs, the matter that was previously constrained by those magnetic fields suddenly finds itself able to expand under the intense pressure. Instead of a single coil springing open as in a flare, it's as if they are all free to expand. The matter suddenly bursts outward in a *coronal mass ejection,* or CME.*

The energy of a CME goes more into accelerating particles than it does into giving off light, so the event is actually difficult to detect initially. In fact, while the first flare was seen almost two hundred years ago, CMEs weren't first seen until the 1970s!

However, their effect is profound. Unlike flares, which are basically a local disturbance, CMEs involve a gigantic area of the Sun. If flares are like tornadoes—local, intense, brief, and focused—CMEs are solar hurricanes. The effect is not as intense, but much, much larger: as much as *a hundred billion tons* of matter are hurled into space at a million miles per hour, and that can do far more damage on a far bigger scale.

As the CME expands off the surface of the Sun, it thunders across interplanetary space and expands to tens of millions of miles across. It creates a vast shock wave as it crosses the thin material previously ejected in the solar wind. It's an interplanetary sonic boom, and it can accelerate subatomic particles to extremely high energy. These particles can gain so much speed that they move at a substantial fraction of the speed of light. It's like a vast tsunami unleashed from the Sun, and it marches outward . . . sometimes toward us.

Once the CME erupts, it can cover the distance from the Sun to the Earth in one to four days. That's all the warning we get.

It's possible to see the actual event when it occurs. When you try to look at an airplane flying near the Sun, what do you do? You put up your hand to block the Sun, allowing you to see the plane. Astronomers do the same thing. They equip sunward-pointing telescopes with coronagraphs—generally very simple masks of metal that block the

* Not all CMEs are associated with flares. Sometimes, they happen all on their own, as more and more field lines get tangled up, resisting the expansion of the matter they constrain. Eventually, the matter breaks free and forms a CME. The magnetic reconnection associated with flares makes it easier to trigger a CME, but it's not always necessary.

On May 13, 2005, the orbiting Solar and Heliospheric Observatory captured this image of a CME heading right for Earth at 3 million miles per hour. When the wave hit, it caused a magnetic storm that spawned aurorae seen as far south as Florida.

SOHO (ESA & NASA)

fierce light coming from the Sun's surface—that allow fainter objects nearby to be seen. When a CME occurs, it can be seen by these telescopes as an expanding puff of light coming out from the Sun. If a CME is seen coming from the side of the Sun, astronomers breathe a sigh of relief: it will miss the Earth because it was aimed sufficiently far away from us. But sometimes the Sun is not so agreeable, and it sends a hundred billion tons of million-degree plasma screaming our way. This is seen as an expanding halo of light, because we are looking down the throat of an advancing front of subatomic particles accelerated to mad speeds.

When it gets here, all hell can break loose.

RINGING THE DOORBELL

The Earth has a magnetic field that is similar in some way to the Sun's. It's probably generated by the motion of hot, molten rock and metal inside the Earth in a process similar to that which takes place in the Sun (with the Sun, though, the material is extremely hot gas), and is powered by a dynamo like the Sun's field as well. This magnetic field extends past the Earth's surface and reaches out into space, forming a region called the *magnetosphere.* If the Earth were alone in space, the field would surround our planet in a shape like that of a doughnut—the three-dimensional version of the crescent-shaped lines seen when you put iron filings on a piece of paper with a bar magnet under it. However, the constant stream of particles flowing past the Earth from the solar wind shapes the Earth's magnetosphere into a teardrop shape, like water forming teardrop-shaped sand banks in a river. The pointy end always faces away from the Sun, and is called the *magnetotail.*

Most people are aware that the Earth's magnetic field can be used to find north,* but it also acts something like a protective force field, rebuffing any passing charged subatomic particle and sending it on its way. This protects us from the more severe effects of solar temper tantrums. It even protects our atmosphere: without the magnetosphere, the solar wind would have long ago eroded our air away, leaving the

---

* In reality, the magnetic north pole and the geographic north pole don't coincide; because of the Earth's ever-changing dynamo the magnetic poles wander, and anyone who needs great accuracy in finding north using a compass needs to correct for that. And things get even worse: what we *call* the Earth's magnetic north pole is actually, by the way magnetic poles are defined, the *south* magnetic pole. It's just by tradition that we call it the north pole. And oh—it gets worse still: the poles on a bar magnet are actually labeled for the pole they attract. So the pole labeled "N" on a bar magnet actually tries to point to another magnet's north pole (it "seeks" the north pole), so the pole labeled "N" is actually the south pole. Confused yet? Yeah, like magnetism isn't already hard enough to understand.

Earth a barren rock similar to Mercury. Mars probably lost most of its atmosphere this way as well.

So the Earth's magnetic field is a good thing. *Usually.*

When a CME from the Sun reaches the Earth, it interacts with the Earth's magnetosphere. The sheer energy of the flow can snap the Earth's sunward-facing magnetic field lines, blowing them back around to the night side of the Earth into the magnetotail, where they can reconnect—it's a bit like a stiff wind blowing your hair backward and making it all tangle up on the back of your head.

When the Earth's field lines reconnect in the magnetotail, a lot of energy is released. Charged subatomic particles flow along the lines, down toward the Earth. Accelerated by the magnetic field, they slam into the Earth's atmosphere, ionizing molecules in the air, stripping them of their electrons. When the electrons recombine with atoms, light is emitted with characteristic colors: oxygen molecules give off red light, and nitrogen green.* Since this happens where the magnetic field lines of the Earth drop down into the atmosphere near the poles, in general people living at extreme northern and southern latitudes who venture outside during such an event are met with a brilliant display of aurorae—*aurora borealis* for the north, and *aurora australis* for the south. In a particularly powerful event, it's possible to see them at mid-latitudes as well; the 1859 white-light solar flare event spawned a massive CME that caused aurorae to be seen as far south as Puerto Rico.

Aurorae have mesmerized people for millennia, and it was only recently understood that they are harbingers of vast unseen forces at play high above our heads, forces that trace their origins back to our nearest star and to the unimaginable violence wreaked there.

The effects of a big CME are far larger than a simple light show, however. For one, they compress the Earth's magnetosphere. A satellite

---

* This is pretty much how a neon sign works; the ionization energy comes from electricity (when the sign is plugged in), and when the electrons recombine with their parent atoms the gas glows. The neon may be mixed with other gases to get different colors.

The Earth is not the only planet affected by the Sun. This ultraviolet image from the Hubble Space Telescope shows an aurora at Saturn's north and south poles. Any planet with a magnetic field can experience magnetic storms when the Sun is active.

J. T. TRAUGER (JET PROPULSION LABORATORY) AND NASA

orbiting above the Earth inside the protective magnetic field may suddenly find itself exposed to the full brunt of the CME. The incoming radiation can then fry it.

There are even more profound effects from a big CME, ones that affect us directly, even on the surface of the Earth.

Remember that a changing magnetic field can induce a current? Well, when the magnetic field of the Earth changes rapidly because of a CME impact, any nearby conductor can suddenly find itself dealing with a huge surge of current.

There are plenty of such conductors on the surface of the Earth . . . like the entire North American power grid. Think of it: millions of miles of wires, all designed specifically to carry current from one place to another! Under normal operating conditions, these wires are easily able to carry a large amount of current, making sure that electricity generated at, say, Hoover Dam can be sent to Los Angeles to power someone's margarita blender.

But these wires are very sensitive to solar storms. For one thing, these storms add a huge load to the system. For another, current heats up wires, causing them to sag. This process is well understood by electrical engineers and under normal operating conditions the system is designed to withstand it. However, a big pulse of current caused by a storm can add to the load already there, causing lines to heat up too much and break. For a third, over the years, more power generators have been added to the grid, but not more wires. As time has gone on, American power demands have grown. The wires were originally built to hold quite a bit of current, but in many cases they are operating closer and closer to their full capacity. A big surge can blow out the huge transformers vital to making sure the high-voltage electrical current in wires gets dropped to much lower voltages before going into your house. These transformers are expensive (some are as big as houses) and losing them can mean whole cities might go without power for great lengths of time.

Case in point: on March 6, 1989, an ugly and enormous group of sunspots rotated into view on the solar surface. Spanning 43,000 miles, they had already spawned many flares that were detected even though the spots themselves were on the far side of the Sun. Astronomers expected the worst.

They got it. Over a two-week period, Active Region 5395 blasted out nearly two hundred solar flares, a quarter of them rating in the highest energy category. At the same time, *thirty-six* CMEs were detected screaming out from the Sun.

Some of the effects were merely annoyances in the grand scheme of things. A microchip manufacturer had to shut down operations temporarily because some sensitive instruments were not behaving during the magnetic upheaval. Compass readings were off by many degrees, making navigation for ships difficult. Many satellites lost altitude—by as much as half a mile—and one military satellite could not compensate for the effects, and began to tumble. Other satellites were fried as well.

But the worst effects occurred on March 13, when a vast geomagnetically induced current was created by the storm. Voltage fluctuations caused power problems around the planet. In New Jersey, the current induced by forces far overhead blew out a power plant's 500,000-volt transformer, which cost $10 million to replace. It took six weeks, and the company lost nearly twice that much money in lost power sales during that time.

In Quebec, the effect was much more serious. The current surge shut down a power generator, and the sudden loss of power collapsed the grid. Transmission wires failed over a huge area, some exploding in flames. In the middle of a winter's night, the electricity for *six million people* in Canada was flicked off by the Sun. It took days to get the grid fully back up. Models of the event made by engineers estimated the total damage cost at several *billion* dollars.

As with asteroid impacts, there are ways to mitigate the damage done by flares and CMEs. Satellites can be designed to withstand particle and gamma-ray impacts, but at a significant cost to the manufacturer. The same is true for power grids; it would cost billions to retrofit power stations and add more power lines to accommodate another March 1989 event.*

Such events are rare, occurring two or three times per century. But as we make more demands on our power grids, the risk of potential damage from the Sun only increases.

---

* Oil and natural gas pipelines are conductors too. An electric flow in a pipeline can increase the corrosion rate of the metal, because the voltage change can increase the ability of moist soil to erode the metal. This won't cause a spectacular collapse like the 1989 power grid failure, but it does reduce the operating lifetime of the pipeline, and can cost billions to retrofit.

And there is yet another direct impact from solar activity. Models of the impact of the 1859 event on our atmosphere have shown that the subatomic particles accelerated in the Earth's magnetosphere by the event would have cascaded down into the atmosphere, breaking up (what scientists call *dissociating*) molecules of ozone in the upper atmosphere. Ozone is a molecule made up of three oxygen atoms (the molecule of oxygen we breathe has two atoms bound together), and is very efficient at absorbing the Sun's ultraviolet light, protecting us from it. The amount of ozone depletion from the 1859 flare would have been relatively modest, just a few percent. However, that *is* enough to allow increased UV radiation to reach the Earth's surface. The effects of this on humans are unclear because of spotty medical records from more than a century ago, but it's possible there was a small but significant rise in skin disease in the years following the event. This increase in UV can also affect the ecosystem and food chain (see chapter 4 for more details on that than you want to know), though again the records from that time are incomplete.

There was, however, at least one measurable effect from the 1859 event. When broken up by incoming particles, the dissociated air molecules can recombine to form other chemical compounds, including NO_2, or nitrogen dioxide.* This reddish-brown gas, created high in the atmosphere, would wash down to Earth in rain and be deposited on the ground. Studies of ice cores from Greenland have shown an increase in the deposition of nitrates from that time.

The problem is, the NO_2 can oxidize in the atmosphere to form nitric acid. When this dissolves in water droplets, acid rain can result, with terrible effects on the Earth's ecosystem. This did not appear to be a major problem from the 1859 event, but if in the future more energetic eruptions impinge on our atmosphere, we may be able to measure the effects of that as well. That's just another fun way the Sun can slam us.

---

* Not to be confused with nitrous oxide, N_2O, or laughing gas; nitrogen dioxide is far more serious.

CLIMATE OF CHANGE

With all this talk of magnetic storms, flares, and CMEs damaging the Earth, are we missing something more obvious? The Sun is, after all, far and away the major source of heat in the solar system. While the Sun seems rock-solid in its energy output, we have already established that it's a variable star. Sunspots wax and wane on an eleven-year cycle; could this possibly lead to a change in the amount of energy we receive from the Sun? And if more or less sunlight hits the Earth, could that then lead to climate change on Earth, and a potential mass extinction?

It should be noted immediately that time and again, people have tried to tie the Sun's eleven-year cycle with events here on Earth. The stock market, baseball scores, even personality traits have been (dubiously at best) linked to sunspot numbers. The problem is, if you look at enough cycles, some are bound to line up superficially. You have to be able to separate the wheat from the chaff, which can be very difficult.

Scientists have been arguing for years over whether there is some correlation between solar activity and weather on Earth. It seems that there is, but the factors involved are subtle and difficult to pin down. If they were clear, there'd be nothing to argue over. However, there are some connections that appear to be firmly in place . . . and sunspots do play a role. But the direction of that role might surprise you.

Sunspots are dark, cooler patches of the Sun's surface. You might think, then, that if there are lots of sunspots, we get less light and therefore less heat from the Sun. So, lots of sunspots equals cooler climates.

But spots are only dark in *visible light.* There are bright regions surrounding sunspots called *faculae* (literally, Latin for "little torches") that form because of the complicated connection between the Sun's surface magnetic field and the hot gas bubbling up from deeper regions. The gas in the faculae is hotter, and therefore brighter. On average, sunspots are 1 percent darker than the Sun's surface, but faculae are 1.1 to 1.5 percent brighter. This means that when the Sun is covered in spots, it's actually *brighter* in visible light than it is when there are fewer spots!

The primary source of heat for the Earth's surface is the visible light from the Sun. Studies have shown that when the Sun is at the peak of its cycle—when sunspots and faculae are more prevalent—the overall solar irradiation of the Earth increases by just about 0.1 percent. This is a small but significant increase—it causes a global temperature increase on Earth of about 0.1 to 0.2 degree Celsius (about 0.2 to 0.4 degree Fahrenheit). The opposite is also true; during the sunspot minimum, the Earth's average temperature decreases by a fraction of a degree.

Let's face it: this is a pretty small effect. By itself, it hardly changes anything on Earth. However, heating of the Earth's surface from the Sun is only *one* way the climate can be affected. There are lots of other sources of climate change, as we are now all too aware. In many cases, these sources *by themselves* don't do much to the climate.

But what if two or more of these effects add up?

Things can get bad. We need only to look back in time a short way to see how.

The existence of sunspots had been known for centuries, even before the invention of the telescope. But once telescopes were trained on the Sun, the view naturally improved. People have been monitoring the size and number of sunspots nearly continuously since the early 1600s.

In 1887, an astronomer named Gustav Spörer noticed that the records of sunspots appeared to show an absence of spots between the years 1645 and 1715. For literally seventy years, the Sun's face was virtually blank, clean of solar acne. In the late 1800s, the scientist E. W. Maunder summarized Spörer's findings and published them. We now call this period of sunspot deficit the Maunder Minimum.

All of this would be somewhat academic if not for one rather critical point: the years 1645 to 1715 were also a time of much lower than average temperatures across Western Europe and North America. It was so cold that the Thames River froze over (which it generally does not do, even in winter), glaciers in the Alps advanced, destroying whole villages, and the Dutch fleet was frozen solid in its harbor. This period was called the Little Ice Age.

It's awfully tempting to directly connect the Maunder Minimum with the Little Ice Age, but we have to be very careful. In nature, it's rare for a single effect to have a single cause, especially when the effect is as dramatic as a prolonged climate change. Usually, there are a number of events that have to occur to manufacture such a big change.

It turns out the Little Ice Age may have started long before the Maunder Minimum, even as early as the mid-thirteenth century. Caspar Ammann, a solar physicist who has extensively studied the connection between the Sun's output and the Earth's climate, notes that the Little Ice Age was not one continuous event, but instead consisted of "several pulses of cooling episodes . . . the first one started in the 1250s through 1300, after a medieval warming period." Clearly, there were other causes of the temperature drop.

The biggest culprit is probably volcanic activity. There are clear signals of eruptions during the Little Ice Age, mostly seen in ice cores: atmospheric gases trapped in polar ice can be studied to determine what was happening in the Earth's air during certain times in history. Interestingly, in the 1690s, the Little Ice Age got very severe, especially in Western Europe—there are stories of birds literally freezing to death sitting in branches. At this very time, there is a large spike in the amount of atmospheric sulfur found in ice cores, indicating large levels of volcanic activity. Volcanoes launch sunlight-reflecting dust and gases into the air, reducing the amount of visible light reaching the Earth's surface. This cools the planet by lowering the amount of heat the surface can absorb.

By itself, this could not cause the severest parts of the Little Ice Age. But together with the Maunder Minimum, when the global temperatures would have dropped, it could have lowered the Earth's average temperature even more.

Still, if this were a *global* effect, why was Western Europe hit so much harder than everywhere else?

It turns out there is a *third* player in this game. This gets a little complicated, so strap yourself in.

During a sunspot minimum, there is less solar activity in general. Besides there being less visible light, there is a drop in the amount of sunlight across the spectrum, including ultraviolet light. This turns out to be important: UV light is what helps create the Earth's ozone layer; it turns normal atmospheric oxygen (O_2) into ozone (O_3). If there is less UV, there is less ozone. Ozone is actually quite important in the temperature balance of the upper part of the atmosphere, called the *stratosphere.* When there is lots of ozone the stratosphere is warm (because it absorbs UV light), and when there is less ozone the stratosphere is cooler.

Most, but not all, of the ozone creation happens in the tropics, at low latitudes near the equator. That's because that's the part of the Earth getting the most sunlight, and therefore the most UV. In the summer, ozone can be created both at the equator and at the pole, because that whole hemisphere is in sunlight. In that case, the difference in temperature in the stratosphere from pole to equator is minimal.

But in the winter, the pole is in darkness. No UV reaches the stratosphere, so no ozone is created there. That in turn means there is a big temperature difference in the ozone layer between the equator and the pole.

The problem is that the jet stream is sensitive to these temperature differences. In the winter, the temperature change across latitudes is large. This drives a strong jet stream, which circulates very firmly around the globe. But in the summer, when the gradient is smaller, the jet stream weakens. Instead of making a tight circle, it meanders, flopping down loosely to lower latitudes. When it does this, it can bring cold air from the Arctic to southern locations, and warm air from the south up to higher latitudes.*

As it happens, the jet stream tends to dip down more at certain locations on the Earth than others. Western Europe is one such place.

This then is the most likely scenario for the very bitter winter cold snap in the 1690s in Europe: volcanic activity dropped the global temperatures, as did the Maunder Minimum. Together they made things

---

* If you live in the northern hemisphere, that is. For Australians, New Zealanders, and other upside-down people, reverse these directions.

cold, but not brutal. But the drop in solar activity dropped the Sun's ultraviolet output, which lowered ozone production on Earth. This changed the direction of the winter jet stream, bringing the unusually cold Arctic air down to Western Europe.

And then people could ice skate on the Thames.

It should be noted that in Western Europe, "the summers were not all that unusual," according to Ammann. This indicates that whatever caused this intense pulse of cold weather was restricted to winter, which is consistent with the above series of events.

Like I said, this is complicated. *But that's the whole point.* If it were simple, we'd understand it better, and no one would be arguing over how the Sun affects the climate. In fact, these events are all fairly well established in general, but the problem is the *magnitude* of each one. How much less ultraviolet light was emitted by the Sun during the Maunder Minimum? How much less ozone was created? How far did the jet stream dip south? How much sulfur was spewed into the air by volcanoes? Changing any one of these inputs makes the results different, so knowing how much each one affects the climate is very difficult to figure out.

The important thing to remember here is that while the Sun affects our climate, changes in its total output over the eleven-year magnetic/sunspot cycle are small. There is a definite effect on the Earth, but it's more like a priming charge than the explosion itself. It requires other catastrophic effects—volcanoes, asteroid impacts, man-made emission of CO_2 and methane—to take advantage of the Earth's climatic sensitivity and cause a disaster.* And even then, at least in this particular

---

* Even very subtle changes in the shape of the Earth's orbit can have an effect here, given enough time (like, millennia). Some scientists even speculate that the Sun's magnetic field protects the Earth from an onslaught of subatomic particles that come from deep space. Called *cosmic rays,* they might seed cloud formation in our air and thus lower the Earth's temperature. However, the data are *very* marginal for this claim. Much more study is needed to understand these effects. Cosmic rays do have deleterious influences on Earth, which we will see in subsequent chapters, but we can't include climate change among them just yet.

case, the problems tend to be regional. The global environment of the Earth doesn't change that much.

That's cold comfort to people who are affected, of course. And if the particular region is very sensitive—or that region has global impact itself—then the results can be much worse. A decades-long series of brutal winters in the United States, for example, or China, could cause famine and economic depression. Wars start over such things, and modern wars can wreak far more damage than a simple solar minimum. When it comes to potential extraterrestrial sources of destruction, the last thing we need to do is add our own capabilities to them.

A more pertinent thought is: could another such minimum occur again? Yes, it could. Worse, it doesn't look as if such events are entirely predictable. Scientists studying the occurrence of long minima in sunspot numbers show that they don't appear at regular intervals, meaning they are not an inherently predictable phenomenon in the long term, although it's marginally possible to make predictions about the very next sequence in the solar cycle. So we might be headed into another minimum a few cycles from now, or it might not happen for a thousand or ten thousand years. But it seems very likely indeed that it *will* happen again.

HOT PLANETS AND HOT AIR

So if the Sun can indeed affect Earth's climate, what about global warming? Is it caused by the Sun, and not by humans?

A lot of noise has been made on this topic, but scientists actually do agree on this: the Sun is *not* the cause of the current temperature rise seen in the latter half of the twentieth century to today.

This isn't hard to show, actually. The amount of radiation from the Sun is measurable, and since the 1950s to today there has not been an increase in solar radiation. In other words, the Sun has *not* been getting brighter during the time when the Earth has been getting warmer. The amount of solar radiation has been quite steady since 1950, and is obviously not the cause of global warming. It's clear to the overwhelming majority of scientists independently studying this phenomenon that it is human activity, *our* activity, that is behind the current sharp rise in global temperatures.

This most basic fact has not stopped some people from claiming that many *other* planets are also experiencing global warming, and therefore the cause here on Earth cannot possibly be human-induced. The only thing linking all the planets is the Sun, they say, and therefore the Sun is causing this warming.

However, this is nonsense. The claim is that Mars, Jupiter, Triton (a moon of Neptune), and even Pluto are warming.* However, each of these has separate causes, linked with the individual objects' atmosphere and orbit, and any purported warming is not related to the Sun.

And let's be clear: these objects are much farther from the Sun than the Earth, and receive proportionately less heat. To warm up Pluto even one degree, the Sun would have to get so much brighter and hotter that it would be overwhelmingly obvious—in fact, the Earth would get totally fried. Since our own warming is less than a degree, it's clear that the other planets' warming must be due to some other source than the Sun.

SUNNY OUTLOOK

We live on a small planet where a considerable number of factors have to align to make life hospitable. However, we live near a tempestuous star that will, inevitably, do what it can to disrupt that equilibrium. Ironically, too much solar activity can cause immediate and global damage, but too little can, in the long run, be just as bad. Like most things in the Universe, this is a delicate balance, and a swing to either side would be catastrophic.

However, we have survived many small oscillations. The Little Ice

---

* I'll note that it's not even clear that these objects are actually warming at all; the data are sparse. In the case of Jupiter, for example, it's not a global effect; only small sections of it are warming because of local atmospheric conditions.

Age came and went, with people taking it in stride—they really did skate on the Thames. Huge flares have wreaked havoc on our power grids, and with a little care, foresight, and a pile of money, we can avoid total disaster.

As for the big swings . . . well, we'll see. They may not happen for centuries or even millennia, and by then we may be able to take action. But the time to start thinking about it all is right now; and we are. Smart people are working on these very topics, and while it may take time to figure out all the angles, and there may be lots of arguments along the way, I think in the end we'll figure a lot of this stuff out.

In the meantime, I'll still enjoy the occasional sunny afternoon . . . but I'll also be mindful that over my shoulder, just an astronomical stone's throw away, is a vast and mighty star. And it has a temper.

CHAPTER 3

The Stellar Fury of Supernovae

THE FIRST ONES TO NOTICE ARE PROFESSIONAL *astronomers.*

Researchers at the Super-Kamiokande neutrino observatory in Japan are shocked when their detectors light up like Christmas trees. Such unprecedented readings prompt them to look for malfunctions in their hardware, because surely no astronomical event could generate so many of the ghostly subatomic particles—even the Sun, the brightest object in the sky, barely produces enough neutrinos to be picked up by their instruments. There would have to have been millions of neutrinos detected to register so strongly! Poring over their instruments, it takes them nearly two hours to figure out that the flood of neutrinos was indeed real, which was far too late . . . not that advance knowledge would have helped.

Within minutes of the event, automated observatories orbiting the Earth perk up. Astronomical satellites designed to observe high-energy light such as X-rays and gamma rays see a rise in detections. One by one, as they slew over to focus on the source of the particles, their detectors saturate with photons as the fierce light intensifies. Within minutes the satellites are blinded, overwhelmed with light, and lose track of the target.

On the ground, across the night sky of Earth, thousands of amateur astronomers, truckers, police officers, and general night owls notice the light in the sky. It's getting brighter by the minute. Some of the amateur astronomers momentarily think it's an airplane, or the glint of reflected sunlight off an orbiting satellite. But many immediately realize what's happening, and start taking data. Others send out e-mails, alerting astronomers all over the world. Get out your scopes! There's a new supernova!

But the e-mails are unnecessary. Within minutes, the "new" star is so bright that other stars in the sky can't compete. Like the sunrise or the full Moon, the supernova is washing out the sky around it.

Astronomers are beside themselves with glee. It's been over four hundred years since the last naked-eye supernova in our galaxy, and this one will no doubt be a record breaker.

But their joy is short-lived. In the middle of their observations, all their machines suddenly lose power. The images and data are all lost when the computers controlling the telescopes die. And before they can properly assess the problem, all the power goes out. One astronomer ventures outside to see what's going on and realizes that the glow of the nearby city is gone. Normally, the combined luminescence of thousands of streetlights, buildings, spotlights, car dealerships, and house lights drowns out the fainter stars in the sky. In a huge ironic twist, the power is out everywhere and the sky is truly dark for the first time in years, yet she cannot observe because her power is off too. Her telescope is useless.

She stares upward at the stars and, after a few minutes, realizes the sky isn't as dark as it was earlier: the fierce eye of the supernova is glaring down on her, and the sky around it is blue. No other nearby star could possibly compete.

Her attention is diverted when she sees another bright light in the sky, this one moving slowly across the ever-bluer sky. She realizes it's the International Space Station. She laughs, glad to see something normal for a moment.

What she doesn't realize is that the astronauts on board are dead. Had she known, she certainly wouldn't have smiled. But then, in a few years, everyone on Earth will be dead too. Humans were doomed from the instant the first rays of light from the supernova touched the atmosphere.

Gamma rays from the supernova destroy vast amounts of ozone, which is quickly reduced to half its normal amount. When the Sun rises in the morning, its ultraviolet light will stream all the way through the atmosphere nearly unabated. Severe sunburn will be the least of the problems faced on Earth as the UV radiation kills off the ocean's phytoplankton, which make up the base of the food chain. Animals that feed off phytoplankton find their food source dwindling and eventually disappearing in mere days, and animals that feed off those *animals face the same dire problem a few days later. This die-off marches up the food chain, and it won't stop until it reaches the top: us.*

It's been a long time since an astronomical event touched off a mass extinction. But now, another one is under way.

A STAR IS BORN

If you go outside on a dark, clear night, far from city lights, you can see thousands of stars sprinkled across the sky. They may seem unchanging, fixed—some people even refer to the night sky as the "starry vault," implying a strength and permanence. Sure, the stars rise and set, but that is a reflection of the Earth's motion, not theirs. They twinkle too, but again the fault lies in ourselves and not the stars—they flicker because the ocean of air above our heads blurs their light.

Even if you go out night after night, week after week, you may not see any changes in the stars. A sharp-eyed observer may note that some stars subtly and periodically change their brightness; these so-called *variable stars* wax and wane over days and weeks. But the stars themselves neither appear nor disappear, and do very much seem as permanent as the night sky itself.

But the Universe is deceiving. Things do change, and sometimes that change can be dramatic. On July 4, 1054, a new star appeared in the sky in the constellation of Taurus. Chinese astronomers recorded this "guest star," noting that it appeared to be even brighter than the planet Venus, which is third only to the Sun and the Moon in our sky. There are records scattered throughout the world of the appearance of this new object, though they are spotty and not without some controversy, but there is no doubt about the reality of the event.

Today, a thousand years later, if you use a pair of binoculars to scan the sky in the constellation of Taurus between the horns of the bull, you might note a faint fuzzy blob that is clearly not a star. A small telescope can back up this observation. A big telescope—especially one equipped with a camera capable of taking long time exposures—reveals this object to be a gaseous and filamentary cloud. It *looks* like the aftermath of an explosion. In fact, images taken many years apart reveal the gas cloud (called a *nebula* by astronomers, from the Latin word for "cloud") to be expanding; filaments and knots in the cloud have obviously moved

The Crab Nebula is the expanding debris from a massive star that exploded in July 1054. It is perhaps the best-studied object in the sky, and one of the most beautiful.

NASA, ESA, J. HESTER AND A. LOLL (ARIZONA STATE UNIVERSITY)

outward in the intervening time. Backtracking the expansion shows that all the gas originally came from one point in the sky, the position of which is marked by a star very near the center of the cloud, indicative of a single explosive event. By measuring the *speed* of the expansion, the *date* of this event can be estimated. Remarkably, that date is the mid-eleventh century, suspiciously close to when the Chinese guest star was observed. Today, no astronomer on the planet doubts that the two events are the very same thing.

What the Chinese saw was one of the largest and scariest events in astronomy: a *supernova.* It might not have seemed all that scary at the time—after all, it was just a light in the sky! But upon closer examination the magnitude and scale of the event are revealed.

The gas cloud—called the Crab Nebula because of its current dubious resemblance in a small telescope to a crustacean—is the expanding remnant of this stellar explosion. In the subsequent millennium since its creation the cloud has grown to trillions of miles in diameter. The gas is still ferociously hot, heated to thousands of degrees by the shock waves generated as it expands supersonically and rams into the cooler gas surrounding it. Energy also continues to be poured into the cloud by the emanations from the central star, the cinder left over from the explosion.

The distance to the Crab is an estimated 6,500 light-years, or about 40 *quadrillion* (40,000,000,000,000,000) miles, and yet even at such a distance, and after ten centuries, it is one of the brightest nebulae in the sky. At the time, the supernova event itself was bright enough to be seen in full daylight, indicating that in just a few weeks the explosion released awesome amounts of energy into space—as much as the Sun will put out *over its entire lifetime of 12 billion years.* In fact, a typical supernova can easily outshine the combined light from all the hundreds of billions of stars in an entire galaxy, and persist that way for weeks.

Our eyes can only see visible light, a very narrow slice of the energy band of light called the electromagnetic spectrum, which includes radio waves, infrared, ultraviolet, X-rays, and super-high-energy gamma rays. If you had X-ray eyes, the Crab would be one of the brightest objects in the sky. The same is true of radio waves, and if you could see in gamma rays, the Crab would be the single brightest persistent object in the sky.

I'll gently remind you that the Crab is *400 million times farther away than the Sun.*

Clearly, supernovae are awesome events capable of wreaking destruction on a vast scale. At its remote distance, the explosion that generated the Crab Nebula was little more than a pretty light in the sky, but not all supernovae are so far away. In fact, the Earth has had close shaves with exploding stars in the past, and there will certainly be more in the future.

But how close is *too* close? To understand just what a supernova can do to its environment and just how these events can be a danger to us

Hubble snapped this picture of Supernova 1994D, the fourth exploding star seen in 1994. The host galaxy is called NGC 4603, and is located a very safe 100 million light-years from Earth. The supernova was about as bright as the entire galaxy.

NASA AND J. NEWMAN (UC BERKELEY)

on Earth, we'll have to understand what would make an otherwise stable star explode.

THE LIFE OF A STAR

While ancient astronomers were baffled by the stars in the night sky—were they holes in the vault of the sky, letting the radiance of the Sun through?—we have a pretty good understanding of them now.

Stars are not just points of light—each is a sun in its own right, most smaller but some fantastically larger and more luminous than our Sun. What a revelation that must have been, the first time someone realized that stars are suns, but terribly far away!

As astronomers studied stars, slowly and inevitably they learned more about them. Some stars are red, and some are blue (you can see this with your own eyes by examining a handful of the brightest ones), which indicates that they have different temperatures: red stars are cooler, blue stars hotter. Many stars are not individuals, but instead are pairs of stars orbiting each other in what are called *binary systems,* only masquerading as single stars because of their remote distance. Using laws of physics established by the astronomer and mathematician Johannes Kepler in the seventeenth century, astronomers could determine the masses of the stars in binaries, opening the door to an understanding of the physical processes inside them.

At the most basic level, a star is a big ball of gas, and so its behavior is in many ways simple. As a gas is compressed its temperature will rise. A ball of gas with the mass of the Sun will compress under its own gravity, heat up, and shine brightly, but it will have a limited lifetime—without an internal source of heat it will cool in about a million years or so.

By the nineteenth century there was mounting evidence the Earth was at least millions of years old, and perhaps even older. And surely the Sun was older than the Earth! Then, in the 1930s, scientists realized that a star is a nuclear furnace, like a vast H-bomb, whose explosion is held in check by the star's own gravity. Nuclear fusion could support the Sun's energy output for not just millions but *billions* of years, solving its age crisis. In an ironic twist, objects as huge as stars are powered by the tiniest of objects: atomic nuclei.

A typical atom is composed of a dense nucleus in its center surrounded by a cloud of negatively charged *electrons.* The nucleus contains electrically neutral *neutrons,* and positively charged *protons.* The number of protons is what gives the atom its characteristic properties: for example, hydrogen has one proton in its nucleus, helium has two, oxygen eight, and iron twenty-six. Under some circumstances (intense heat or absorption of ultraviolet light, for example) electrons can be removed from an atom, but it's the number of protons in the nucleus that defines the atom.

As you might remember from grade school science, like charges repel. If you try to squeeze two atomic nuclei together, their mutual positive charges resist it. But in the cores of stars, temperatures are in the millions of degrees—meaning the atomic nuclei are zipping along very quickly, making collisions between them frequent and violent—and pressures are so high that the nuclei are squeezed together very hard indeed. If that electrostatic repulsion can be overcome, other nuclear forces take over that merge the nuclei, fusing them together.

This nuclear fusion does two things. First, it creates a new type of atom, since the new nucleus has more protons than either of the two nuclei before the merger. In general, four hydrogen atoms fuse together to make helium (two of the hydrogen protons become neutrons in the new helium nucleus), three heliums fuse to make carbon, and so on. The actual process is far more complicated than this, but that's the basic idea.

Just as important, nuclear fusion releases energy. When you look at the overall process of fusing nuclei, you would expect that the total mass of the atom created by fusion would equal the sum of the masses of the atoms going into the process—a lump of clay created by smacking together two smaller lumps of clay would have the same mass as the sum of the two lumps, of course. But nuclear physics is different from what we see in the everyday macroscopic world: atoms are ruled by quantum mechanics, with its weird properties and common-sense-defying behaviors.

In the process of nuclear fusion, a small amount of the mass is converted into energy. The energy produced is *enormous* compared to the mass; it follows Einstein's famous equation $E = mc^2$, where the energy produced is equal to the mass times the speed of light squared—and the speed of light is a very big number. Even so, the mass converted is so tiny per atom that the energy released is incredibly small—it would take a million hydrogen atoms fusing into helium to equal the energy released when a flea jumps.

But stars are vast repositories of hydrogen. As we saw in chapter 2, in the core of the Sun, 700 million tons of hydrogen are fused into 695 million tons of helium *every second*! The missing 5 million tons are converted to energy, and that is enough to power the star, letting it give off the heat and light we need to survive. In fact, the heat released is what holds the star up from its own gravity: the pressure to expand outward from the energy release balances the gravity trying to crush the star. An equilibrium is maintained as long as the gravity and energy release remain constant.

As stars go, the Sun is on the big end of the scale (most stars are much less massive, less energetic, and less luminous); however, far larger and more massive stars exist. The nuclear fusion in a stellar core depends very strongly on the mass of the star, with the rate increasing rapidly with mass. A star with twice the mass of the Sun fuses hydrogen into helium in its core more than ten times faster than the Sun does, and is therefore ten times as luminous. A star with twenty times the mass of the Sun—and many such stars exist—"burns" its nuclear fuel *over 36,000 times faster* than the Sun. Even though such stars have more fuel, they go through it so much more quickly that their lifetimes are significantly shorter; the Sun will fuse hydrogen steadily for billions of years, while a 20-solar-mass star might live only a few million.

They say that even the brightest star won't shine forever. But in fact, the brightest star would live the shortest amount of time. Feel free to extract whatever life lesson you want from that.

What happens when the hydrogen runs out? It should be noted that a star like the Sun never *really* runs out of hydrogen; most of the mass of the star, in fact, *is* hydrogen! But fusion only occurs in the core, where the pressure and temperature are highest. In the outer layers of a star it is much cooler (tens of thousands of degrees as opposed to millions), so fusion cannot take place. This gas isn't available to the core anyway, so it can't be fused. It's like having a gas can in the backseat of your car. It's there, but it doesn't do much good while you're driving.

But in the core, eventually, the available hydrogen runs low. As the process of converting hydrogen into helium goes on, the helium nuclei build up in the very center of the star. Because helium has two protons, its nuclei resist coming together even more than hydrogen nuclei do, so it takes higher temperatures and pressures to fuse them. For stars with half the mass of the Sun or less, these conditions are never met. Eventually the star runs out of available fuel, and energy generation ceases.

But for more massive stars the helium "ash" can continue to build up. The core gets more massive, its own gravity crushes it more and more, and eventually the conditions for helium fusion are met. In a flash, helium nuclei smash together to form both carbon and oxygen nuclei. This process releases even more energy than hydrogen fusion, so the star becomes more luminous—it literally gets brighter. All the extra heat from the core is dumped into the surrounding envelope of hydrogen. This throws off the balance of pressure outward versus gravity inward, so the star responds as any gas does when heated: it expands. The star swells in size to epic proportions.

Ironically, though, the outer layers of the star cool off! While the total energy emitted by the surface of the star increases, the surface area increases even more. Each square inch of star emits less energy; it's just that there are a *whole* lot more square inches than before. Even though the star gets more luminous, it cools off, becoming red. Because of its color and size, the star is called a *red giant.*

This is the eventual fate of the Sun. Eventually carbon and oxygen

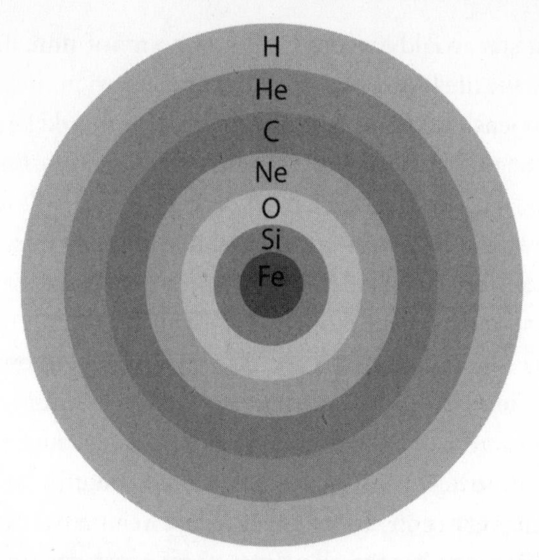

Just before a massive star explodes as a supernova, elements are piled up in its core like the layers of an onion. Iron sits at the very center, surrounded by shells of silicon, oxygen, neon, carbon, helium, and hydrogen. When a star gets to this stage, it doesn't have very long to live.

AURORE SIMONNET AND THE SONOMA STATE UNIVERSITY EDUCATION AND PUBLIC OUTREACH GROUP

will build up in its core, and just as before, it takes more heat and pressure to fuse them than helium. The Sun doesn't have what it takes to fuse carbon or oxygen, and the process ends there.*

Stars with more than about twice the Sun's mass *do* have what it takes to get to this third round of nuclear fusion. In their cores carbon can fuse into neon, releasing even more energy. But it takes even more massive stars to get neon to fuse into magnesium and oxygen, and more massive stars yet to get oxygen to fuse to silicon.†

Silicon will fuse into iron, but it takes a vast amount of pressure and

---

* Of course, the devil's in the details. See chapter 7 for an account of what happens to the Sun next.

† In reality there are many fusion processes going on, sometimes simultaneously. For example, neon fusion produces the heavier element magnesium and the lighter element oxygen. Then oxygen fusion produces silicon, sulfur, and phosphorus. However, I decided to simplify all this to avoid a profusion of fusion confusion.

heat, and that can only come from stars with a mass more than twenty times that of the Sun. All of those steps get their turn in such a star, one after another. Each of the steps in the chain, though, takes less and less time, since the temperatures and therefore the fusion reaction rates increase hugely with each process. A 20-solar-mass star will fuse hydrogen for many millions of years, helium for one million years, carbon for a millennium, and neon for just one short year (those steps happen even more rapidly for stars with more mass).

Eventually, the massive star's core is layered like an onion: hydrogen lies in a shell on the outside, surrounding a shell of helium, surrounding a shell of carbon, then neon, then oxygen, then silicon. Finally, at the deepest part of the core is a sphere of white-hot iron. To be sure, there is some mixing going on, but in general the layers are fairly well separated. But this is just the core: the outer layers of the star up to the surface are still almost entirely nonfusing hydrogen. These layers absorb all this heat being created in the core, and, as in their less massive cousins, this gas swells out, becoming grossly extended. Stars in this mass range, though, get far larger than red giants. They can swell to diameters reaching many hundreds of millions of miles, and so we call these bloated beasts *red supergiants.*

In such a massive star, after millions of years, the fusion cycle is nearing its end. Iron is different from other elements. Unlike hydrogen, helium, and the others, iron resists fusion under almost any circumstances. No normal star in the Universe can produce the temperature and pressure needed to fuse it. At the very heart of the star, deep inside its core, a ball of inert iron just a few thousand miles across sits there ticking like a time bomb. And when enough of it builds up from silicon fusion, the bomb goes off.

RAGE, RAGE INTO THE NIGHT

And now, finally, we have come to the moment of truth. For a year the iron has been accumulating in the massive star's core, and all that time has been writing the star's death sentence.

Until this point in the star's life the core has been generating energy; now this has stopped. Remember, the heat from nuclear fusion is one factor that supports the star against its own crushing gravity.

A second source of support against gravity is the tremendous sea of electrons in the core. In a normal atom, electrons stay connected to the nucleus. However, in the core of a star the conditions are so extreme that electrons are stripped off their atoms. Anytime an electron tries to attach itself to a nucleus, the intense heat and pressure rip it off again.

In the core, electrons are squeezed together very tightly, and weird quantum mechanical effects become important. One of them is called *degeneracy,* which is similar to electromagnetic repulsion: if you try to squeeze too many of the same kinds of particles together (regardless of charge), they resist it. This resistance is a major source of support for the core. Degeneracy, together with the raw heat from nuclear fusion, keeps the star's core from collapsing under its own gravity.

The problem is, degeneracy pressure can only withstand so much gravity. As the iron piles up, the core gets more and more massive, and its gravity gets larger and larger. There is a moment when the iron core reaches its tipping point, when its mass is about 1.4 times the Sun's. At that point, degeneracy loses. It simply cannot hold back all that mass. Previously, when the star was fusing other, lighter elements, this point was never reached; the next element up the chain would start fusing and the star's core was saved.

But iron won't fuse, and degeneracy is no longer enough. The core cannot withstand its own titanic gravity, and its support mechanism fails. Catastrophically. The core collapses . . . but this is no gradual deflation, like a balloon losing its air. When the core of a massive star collapses, it *collapses.* And all hell breaks loose.

The collapse is incredibly fast: in a thousandth of a second—literally, faster than the blink of an eye—the tremendous gravity of the core shrinks it down from thousands of miles across to a ball of ultracompressed matter just a few miles in diameter. The speed of the collapse is breathtaking: the matter falls at speeds upward of 45,000 miles per *second.* The core heats up almost beyond belief, to a billion degrees.

High-energy gamma rays are produced, and these vicious photons are so energetic they can actually destroy atomic nuclei when they collide with them. This process, called *photodissociation*, rapidly starts destroying the iron nuclei in the core, blasting them into bits of helium nuclei and free neutrons. This actually makes things worse (if you can imagine), since these can absorb even more energy, accelerating the collapse.

The events in the core reverberate throughout the star. The core was supporting the outer layers of the star, and when the core collapses, for them it's a real-life Wile E. Coyote moment: just as when the cartoon character suddenly realizes he is no longer over solid ground and starts to fall, the gas from the star's outer layers suddenly finds itself hovering over a vacuum and comes crashing down. The incredible gravity of the core accelerates the gas hugely, and it slams into the compressed core at a significant fraction of the speed of light.

This creates a *huge* rebound effect that reverses the direction of the inbound gas and starts to blow it back out. This rebound, as vast as it is, is amazingly not enough on its own to blow up the star; the explosion would stall, and the outer layers would begin to fall once again onto the core. But the star has one more surprise up its sleeve.

Even after the initial collapse, the core is still loaded with electrons. The tremendous heat and pressure from the collapse applies a huge force on these electrons, squeezing them together into the protons in the core. When this happens, the electrons plus protons create more neutrons. But they also create ghostly subatomic particles called *neutrinos*, and these are what spell disaster for the star.*

Neutrinos are extremely tenuous particles, able to penetrate huge amounts of material without getting absorbed; to them even the densest material is nearly transparent. They blast out of the core, carrying away vast amounts of energy from the collapse. The energy they carry out is nothing short of staggering: it can equal the Sun's *entire lifetime*

---

* The word *disaster* comes from the Latin for "bad star," so in this case we can be literal.

output of energy! In fact, the solid majority of the energy released in a supernova event is in the form of neutrinos; the visible light we see, blinding though it is, only adds up to a paltry 1 percent of the released energy.

The core generates neutrinos in unbelievably prodigious quantities: some 10^{58} (that's a 1 followed by 58 zeros, folks) of the particles scream out of the core over the course of about ten seconds. This is just around the same time that the outer layers of the star fall onto the core and begin their failed rebound. Just as the gigantic bounce fails, and all that material is about to fall back on the core, all those countless neutrinos slam into the gas.

Even though neutrinos tend to pass right through normal matter, the stellar gas is incredibly dense. Plus, there are just simply so many neutrinos that some fraction of them get absorbed no matter what—it's like driving through a swarm of bugs in your car; no matter how much they avoid you, you're still going to get some goo on your windshield.

Only a tiny fraction, maybe 1 percent, of the neutrinos get absorbed by the gas, but it's still an epic event: the total energy dumped into the gas is *huge*.

This, *this* is what destroys the star.

It's like setting off a bomb in a fireworks factory. The energy of a hundred billion *billion* Suns rips into the star's outer layers, reversing their course, literally exploding them outward. Octillions of tons of doomed star tear outward at speeds of many thousands of miles per second. The event is so titanic that even the tiny fraction of it that is converted into light can be seen clear across the visible Universe.

And that's just visible light. Other forms of light—X-rays, gamma rays, and ultraviolet light—also pour out of the newly formed *supernova*. As the shock wave of the explosion tears through the outer layers of the star, pressures and temperatures get so high that nuclear fusion can be triggered. In fact, elements heavier than iron can finally be created in this way, since the conditions in the blast wave are, incredibly, actually *more* violent than in the core of a star. Radioactive versions of elements like cobalt, aluminum, and titanium are created in the expanding debris, and they emit gamma rays when they decay. The gas, already hellishly hot, absorbs this energy and becomes even hotter, heated to millions of degrees. It glows in X-rays and ultraviolet light. Also, these explosions are rarely perfectly smooth. Some materials will be accelerated faster than others, and the inevitable collisions between them generate even more tremendous shock waves, similar to sonic booms inside the expanding material. This can also generate X-rays.

All in all, a supernova is a seething cauldron of power, chaos, and violence. It is one of the most terrifying events in the visible Universe.

ROUGH NEIGHBORHOOD

Needless to say, anything close to the exploding star is facing upwind in a flaming hurricane. Any planet orbiting the nascent supernova is a goner: having your primary star explode in a billion-degree conflagration can end in only one way, and it's not pretty. The planets will be torched, sterilized, and any air or water is stripped away by the sheer energy of the explosion.

The sudden decrease in mass of the star weakens its gravity severely, thus ejecting any planets from the system. It's possible that there are thousands or even millions of scorched rogue planets wandering the Milky Way, their birth stars long since dead. Space is so vast, however, that we may never find such planets even if the galaxy is loaded with them.

Clearly, supernovae are *dangerous.* Your best bet is to stay as far away from them as possible. But *how* far away? If a star in our galaxy explodes, how close is *too* close?

In the appendix is a table that lists all the known stars within 1,000 light-years that have the potential to go supernova. The closest, Spica, a blue giant in Virgo, is about 260 light-years away, and most of the others are considerably farther off. While we can't give the specific date any one of these stars will explode, it is a dead cold fact that they all *will* blow up, and some in the next few thousand years.

How much should we worry about this?

It depends on what it is we should be worried about, actually. At first glance, you might think that just the sheer enormity of the event is all you need to consider. An entire star just exploded! But in fact there are many weapons in a supernova's arsenal. Some are not cause for concern. But others . . .

Kinetic impact

If you're standing near an explosion, the most obvious danger is from debris. That's bad enough if you're near, say, a grenade, but a supernova takes this quite a bit further: the launch of a few octillion tons of gas into space at a significant fraction of the speed of light sounds more than a little dangerous. And it is! But only if you're relatively close by. A planet circling the doomed star is itself doomed, of course, but what if you're watching from the cheap seats, around another star?

To simplify the situation somewhat, let's imagine that all that matter is ejected from the supernova in one instant. We'd see a thin shell of gas expanding outward, its diameter increasing with time. Almost all the mass of the original star is in that shell (the outer layers that explode outward may outweigh the core by several times). As it expands, the area of the shell increases, and so the amount of mass in a given area decreases—it's very much like light emitted from a lightbulb; the farther you are from it, the more the light gets spread out and the dimmer it appears.

The debris from a supernova spreads out too. If you are on a planet near the explosion, more matter will slam into you than if you're farther away. In this case, the amount of impacting material will drop with the square of your distance: if you double your distance, you'll get one-quarter as much material hitting you. But how far away is far away *enough*?

Just to assume a worst-case scenario, let's take an improbably close distance of 10 light-years for the supernova. That means it would be about 60 trillion miles from the Earth.* Let's further assume the total ejected mass is 20 times the mass of the Sun, about typical for your run-of-the-mill supernova. In this case, the amount of matter that would hit the Earth is about 40 million tons.

Yikes! Duck!

But just how much is that really?

That sounds like a lot of material, but it really isn't; a small hill about 1,200 feet tall would have about that much mass. If that hit all at once it would be bad—chapter 1 made that very clear—but remember, this would be spread out over the surface area of the entire Earth. In fact, it's far less than an ounce per square foot over the whole Earth: once spread out, it's more like a single raindrop falling in your backyard.

And we know it wouldn't be an extinction-level event, since we've survived asteroid impacts of this size and larger before. We might notice a slight diminution of sunlight, but no real long-term effects.

We have a more realistic situation, with the explosion of the star in 1054 that formed the Crab Nebula. At 6,500 light-years away, how much debris will impact the Earth? It turns out to be about 100 tons.† And again, while 100 tons sounds like a lot, the Earth gets hit by 20 to 40 tons of meteoric material a day. Debris from the Crab is just a bump on top of our normal daily influx. But you needn't worry anyway: at typical ejection speeds of one-twentieth to one-tenth the speed of light, it will take 100,000 years for that material to hit us—and the event was only 1,000 years ago. Not only that, but the material will certainly never reach us anyway: gas and dust between the stars will slow down and stop the Crab ejecta before it ever gets close.

---

* Just to be clear, there are no stars anywhere near this close capable of blowing up.

† That answer shocked me. I had to calculate it twice to make sure I didn't make some dumb mistake. The Crab is *40 quadrillion miles away,* yet it hurled so much matter outward that 200,000 pounds would hit us even from that distance! Supernovae are immense.

Optical light

Another obvious feature of supernovae is that they're *bright*. The Crab was about as bright as the planet Venus, even from 6,500 light-years away. How close would a supernova have to be for the light to be too bright?

We have to think for a moment about what "too bright" means. Some animals, for example, time their cycles to the Moon. Breeding, feeding, hunting, and so on are timed or at least aided by lunar light. Having a supernova as bright as the Moon hanging in the sky day and night could in theory affect some species.

For a supernova to get that bright, it would have to be at a distance of about 500 light-years. There are in fact one or two stars that close that could explode, notably again the blue giant Spica in Virgo. If it blew up, it would be easily visible in broad daylight, and at night would rival the Moon in the sky, bright enough to read by and to cast sharp shadows! But this extra light would be more of an annoyance than anything else. Bright as it is, the supernova would still just be a point of light in the sky, difficult to look at directly without making your eyes water. However, there wouldn't be any actual physiological damage to your eyes. You'd just learn to avoid looking at it (or wear sunglasses at night).

There would be no added heat from this new source of light; the supernova would still be too far away to actually warm us up. Think of it this way: the Moon doesn't heat the Earth noticeably, so a supernova as bright as the Moon wouldn't either.

One possible problem would be the disruption of some animal cycles, but the effects of this are hard to determine. They might very well be minimal, since even the fury of a supernova dies down with time. Within a few months the explosion will have faded to more tolerable levels. Animal cycles timed with the Moon may be disturbed, but likely would recover.

It's worth noting that the closer a supernova is, the brighter it is. To get as bright as the Sun, it would have to be *much* closer: about a light-year. Not only are there no stars that close capable of exploding, there are no stars that close to us *at all* (except, of course, the Sun itself).

Neutrinos

What about all those neutrinos, created when electrons in the core of the star merged with protons to form neutrons? The total energy emitted is *huge*. Are we in danger from that?

The answer is a little bit difficult to ascertain, actually. Physically, the direct absorption of the energy of a neutrino by a human cell is not terribly worrisome. Neutrinos are incredibly slippery; in fact, just while you are reading this sentence several trillion neutrinos have passed right through your body, and odds are very high that not a single one was absorbed by you. A supernova would have to be impossibly close—as close as the Sun is to the Earth—to be able to directly kill a human being through neutrino absorption.

But before you sigh in relief, there's more to consider. Neutrinos can bounce off the nuclei in atoms, and deposit their energy that way, rather like hitting a bell with a hammer. It turns out that this method of depositing energy is more efficient—that is, more likely to have an effect. If a neutrino did this, a cell nucleus (specifically the DNA there) could be damaged, potentially leading to the development of cancer.

Once again, the exact danger from this is difficult to calculate, but mathematical simulations have shown that a supernova would have to be improbably close to do any damage in this manner. The effects are minimal for a supernova farther away than about 30 light-years, and again it's worth noting that there are no potential supernovae this close to Earth. Your DNA is safe.

Direct exposure to gamma rays and X-rays

Things get stickier when we consider other forms of light. You're almost certainly familiar with X-rays from visits to the dentist's office, or if you've ever broken a bone. Medically, X-rays are wonderful because they can penetrate the soft tissue of our skin and muscles; as far as an X-ray photon is concerned, those cells are transparent. But bones are denser, and more likely to absorb the X-ray. If you put film underneath an arm, X-rays will pass right through soft tissue and expose the film, while bones absorb the X-rays, leaving only a shadow on the film.

However, soft tissue does absorb *some* X-rays, and that's part of the danger. If a cell absorbs the high-energy X-ray, it's like shooting a bullet into an egg. The energy released as the tissue absorbs the energy can destroy the cell. Low-energy X-rays can also damage DNA, potentially causing a cell to become cancerous. While this sounds alarming, it should be noted that a typical medical X-ray procedure is quite safe—Space Shuttle astronauts, for example, who stay in space for two weeks receive a dose of radiation from the Sun equivalent to about fifty medical X-rays with no ill effects. Digital technology has made it possible to lower the dose even more, since digital detectors are far more sensitive to X-rays than film.

Supernovae are a bit brighter than the dentist's X-ray machine, though. However, the X-rays from an exploding star can only hurt you if they can *reach* you. As it happens, we have a built-in shield.

You're sitting in it.

The Earth's atmosphere is very good at absorbing these types of light. Many astronomical sources emit X-rays, but astronomers didn't even know about them until the 1960s because of the Earth's atmospheric absorption. X-rays are blocked while still high in the atmosphere, so they never reach the ground, and even mountaintop telescopes can't detect them. It wasn't until the advent of the Space Age that it was found that stars, galaxies, and other objects emit X-rays.

So we here on Earth are pretty safe from exposure. X-rays, even from a nearby supernova, are absorbed by our atmosphere, posing little threat. But what about any humans *above* the atmosphere? Astronauts orbiting the Earth in the International Space Station are in fact at risk.

Given typical X-ray emission from a supernova explosion, the astronauts will receive a lethal dose if the star is closer than about 3,000 light-years or so. That's quite a long way! There are *many* stars capable of exploding within that distance of us. Astronauts are clearly our most serious casualties from the prompt (that is, immediate) radiation from a supernova.

Gamma rays, which are higher-energy than X-rays, have pretty much the same story. They are absorbed by our atmosphere, and pose little threat to human tissue for landlubbers. However, they actually make things worse for our spacebound crew. The absorption of the gamma ray by a piece of metal—say, the hull of a space station—can lead to the metal emitting many X-rays in response; it's like electromagnetic shrapnel. A solar flare (as discussed in chapter 2) can generate enough gamma rays to do serious harm, and a supernova within a few thousand light-years can still generate enough gamma rays to equal or surpass the amount created in a big solar flare. Direct exposure to these gamma rays can be lethal. The "secondary radiation" from metal absorption can also be very high, lethal in its own right to unprotected astronauts.

Don't forget that our satellites are also sensitive to this event (see chapter 2). Not only that, but the flash of gamma rays and X-rays from a nearby supernova would ionize the upper atmosphere, creating a cascade of subatomic particles. This would create a strong pulse of magnetic energy that can damage our power grid in the same way a solar coronal mass ejection can (see chapter 2 for details on this kind of event). Communications, television, global positioning, high-flying aircraft, and even the supply of electricity by power lines could be severely damaged by this pulse of supernova radiation.

Again, there are several stars ready to pop within that distance. The odds of any one blowing in the near future are incredibly low, but we are now a spacefaring race, and highly dependent on our orbiting infrastructure. The good news is that if governments take the threat from solar outbursts seriously and fortify our infrastructure against that, we'll be safe from supernovae as well.

At least, safe from *that* particular threat. We're not done touring the arsenal quite yet.

Gamma and X-rays, redux

Before you start to breathe too easily, sitting under this ocean of air, you should realize we're forgetting something. It's true that we ground-based humans are safe from direct exposure to high-energy radiation because the atmosphere absorbs this radiation. But then it's fair to ask, *how does this affect the atmosphere itself?*

This is potentially the greatest threat a supernova poses.

Our atmosphere is a many-layered thing. We sit at the bottom, where there's plenty of oxygen mixed in with nitrogen, as well as traces of other gases like carbon dioxide and argon. But up higher, things are different.

As covered in chapter 2, between heights of about 10 to 30 miles above the Earth's surface sits a layer of ozone, which absorbs dangerous ultraviolet (UV) radiation from the Sun. Unimpeded, this UV light would reach the ground and do all sorts of damage, including causing sunburn and skin cancer in humans. Moreover, many protozoa and bacteria, the basis of the food chain on the planet, are very sensitive to UV.

Obviously, the ozone layer is critically important to life on Earth, and as far as a supernova is concerned, it has a big fat bull's-eye painted on it.

When the X-rays and gamma rays from a supernova hit the Earth's atmosphere, they can destroy ozone molecules, leading to the cascading series of events described at the beginning of this chapter. The critical factor, as it has been all along, is *distance.* How close can a supernova be before it damages the ozone layer enough to affect life on the surface?

This is an important issue, and many scientists have taken it very seriously indeed. Some have set up computer simulations to see how much damage a nearby supernova can inflict on our atmosphere. They used a mathematical model of the atmosphere, which includes such effects as the height of the supernova over the horizon, the time of year, the distance, and so on.

Different models yield different answers, but the result seems to be good for us: a supernova would have to be at most 100 light-years away before there would be enough damage to the ozone layer to kill off the base of the food chain. Some models indicate it would have to be even closer, perhaps 25 light-years.

There are no massive stars ready to explode that are that close, so we once again appear to be safe . . . or do we?

SIRIUS DANGER?

I have some more bad news: massive stars are not the only kind that can explode. Low-mass stars like the Sun lack the mass to create the conditions needed for a core collapse, but it turns out core collapse is not the only way to blow up a star.

In a massive star, helium piles up in the core and eventually will fuse into carbon and oxygen. But in a low-mass star, that doesn't happen: there just isn't enough pressure from the weight of the overlying layers in the star to get the helium nuclei to fuse. Instead, helium just accumulates in the very center of the star, forming a dense ball. This helium sphere is degenerate; degeneracy is that weird quantum mechanical state discussed earlier that occurs when too many particles—in this case, electrons—of one type are squeezed together very tightly. As more helium piles on, the degeneracy increases, and the temperature soars (though in this case still not enough to actually fuse the helium into carbon and oxygen).

As we also saw earlier, the low-mass star expands and cools, becoming a red giant. If it's massive enough it might yet fuse helium into carbon, with carbon eventually building up and the cycle repeating. If the star doesn't have the mass to fuse carbon, the fusion process ends there.

But the red-giant star's life is not quite over just yet. While all this is going on deep in the core, at its surface the situation is different. The star's vastly increased size means that gravity at the surface is much lower; the gas there is not held on as tightly as it was before. Remember too that the star's brightness has increased greatly. Any gas particle at the surface is bombarded with light coming up from below. The gas absorbs this light, which gives it a kick upward. That kick can easily overcome the weakened gravity, giving the gas enough momentum to break free of the surface and be launched out into space.

A dense stream of material is emitted from the star. Astronomers call this a *stellar wind*, like a solar wind on steroids. Red-giant winds can be very dense, blowing off thousands of times as much gas as the star did before its core became degenerate; the stream can be so thick that the star's outer layers can be entirely blown off in just a few thousand years. In just a short time compared to the star's life span, it loses as much as half its mass.

When this happens, the degenerate core is eventually exposed to space, and is called a *white dwarf*. Although it can contain the mass of an entire star, it is so dense that it's no bigger than the Earth. The surface gravity is unimaginable, hundreds of thousands of times stronger than the Earth's. A cubic inch of white-dwarf material would have a mass of several *tons*, like compressing dozens of cars into the size of a sugarcube. It's also very hot, glowing at a temperature of over 100,000 degrees Fahrenheit.

After the outer layers are shed in the stellar wind, this ball of ultra-compressed superhot material is now sitting in the center of an expanding cloud of gas. The white dwarf is so hot that it emits a flood of ultraviolet light that energizes the gas in the expanding wind, setting it aglow. Seen from Earth, these gas clouds look like pale, ghostly disks, glowing a characteristic green color due to oxygen in the gas. Astronomers named them *planetary nebulae* because of their resemblance to distant planets seen through the eyepiece, but that's a misnomer: they are the dying gasps of medium-mass stars, and someday the Sun will go through this stage as well (making life here very uncomfortable, so you just know there's a whole chapter later on devoted to this).

From there on out, though, the star's life is rather boring. Eventually the gas expands away, dissipating entirely and mixing with the lonely cold gas that exists between stars. Over billions of years the white dwarf cools, dims, and simply fades away.

But for some white dwarfs, the story does not end there.

Something like half of all the stars in the sky are a part of binary or multiple-star systems: stars that orbit each other because of their mutual gravity. Imagine now such a binary star, with two stars in mutual orbit. Both have roughly the mass of the Sun. One ages somewhat faster than the other; perhaps it is slightly more massive than its companion, and so fusion progresses a bit more quickly. It becomes a red giant, blows off its outer layers, and becomes a dense helium white dwarf.

Eventually, the other star begins to go through the same process. But when it expands into a red giant, its partner star is already a white dwarf, with its commensurate strong gravity. If the dwarf is close enough to this new red giant, its intense gravitational pull can essentially draw material off the other star, literally feeding off it. This gas, which is almost entirely hydrogen, then falls on the surface of the white dwarf and accumulates like snow on the ground.

Things get dicey from there. The gravity of the white dwarf is incredibly strong, squeezing the mass accumulating on its surface immensely. If the mass is raining down too quickly, it piles up on one spot, and the pressure builds there beyond the breaking point. The hydrogen in the pile fuses in a flash, detonating like a thermonuclear bomb—except one with 100,000 times the energy output of the entire Sun.

There is an immense flash, and the accumulated matter blows off the surface of the star despite the intense gravity. Like belching after eating too much food too quickly, this takes the pressure off the white dwarf, and after things settle down, the matter begins to accumulate again, resetting the cycle.

The energy released is gigantic on a human scale, but still much smaller than a supernova, and this event is called simply a *nova.* The white dwarf is largely unaffected by the blast—the amount of matter blown off in the event is only a few hundred times the mass of the

A white-dwarf star greedily sucks down a stream of matter from its companion, a normal star. When enough matter piles up the white dwarf will either erupt as a nova or detonate utterly as a Type I supernova.

DAVID A. HARDY (WWW.ASTROART.ORG) & STFC

Earth,* which is far, far less than the mass of the star—and therefore the cycle can repeat as long as the red giant feeds the white dwarf.

However, if the red-giant matter stream is on the slower side, things are very different. The gas won't pile up as quickly and explode in one spot. Instead, it will get spread out over the entire surface of the white dwarf, forming a shell of inert hydrogen all around it. But this time there is no pressure release, no ability to burp. Since the matter is spread out, the pressure is lower than in the earlier case, and the material continues to build up, getting thicker and thicker everywhere on the white dwarf's surface. Eventually, however, when enough matter piles up, it will still reach that fusion flashpoint.

In this case, it's not just the hydrogen in one small spot that fuses in a thermonuclear flash; it's *all the matter over the entire surface of the star.* The explosive energy released is much, much larger, and eats its way

---

* Note the use of the word *only.* Astronomy has a tendency to crush our sense of scale into dust. The mass of the Earth may seem huge to us, but it's only about a millionth of the mass of the white dwarf.

down into the white dwarf as well as up into space. The energy release is so titanic that it can disrupt the structure of the star itself, causing a catastrophe on an epic scale. The entire star explodes like one enormous thermonuclear bomb the size of Planet Earth. It is literally a disaster: the star goes supernova.

By a cosmic coincidence, the total energy released in this event (called a Type I supernova) is very similar to the energy emitted by a massive star going supernova (called a Type II), even though the two physical processes are completely different. In fact they look so similar that it took astronomers quite some time to figure out that the two events were actually entirely separate. But both release huge amounts of energy, and both are very dangerous if they happen too close to us.

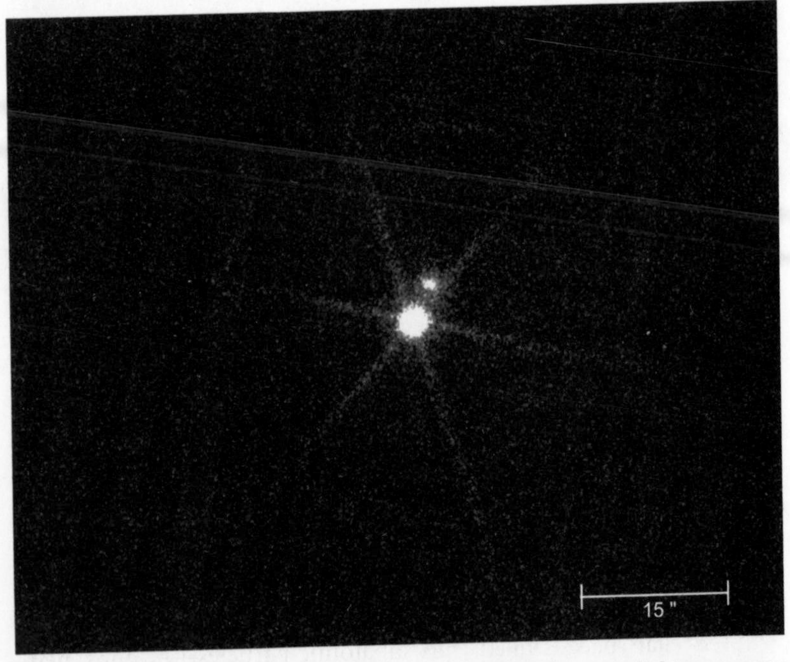

The brightest star in the night sky is Sirius, which is less than nine light-years away. Sirius is a binary, consisting of a normal star like the Sun (but more massive) plus a white dwarf. In this X-ray image, the white dwarf is brighter because it is far hotter than its normal companion. In optical light it is far fainter than the normal star.

NASA/CXC/SAO

One category in which the two events are quite different is their emission of high-energy light: a Type I gives off far more X-rays and gamma rays than a Type II. This means it can be farther away and still hurt us. We know there are no nearby Type II candidates. What about Type I?

Happily, no, none are nearby. However—and there's always a "however"—there *is* a binary star with a white dwarf that is extremely close to the Earth: Sirius, the brightest star in the night sky. It's a mere nine light-years from Earth, which in cosmic terms means it is practically sitting in our lap.

Sirius A, the primary star, is a normal star (that is, fusing hydrogen to helium in its core as the Sun does) with about twice the Sun's mass. In orbit around it is Sirius B, a white dwarf with roughly the same mass as the Sun. Someday, Sirius A will become a red giant, and Sirius B will feed off it . . . but as far as we can tell, Sirius B is way too far from A to be able to feed at the right rate to explode. The white dwarf will indeed accrete matter, and this will cause it to become brighter as the matter heats up and impacts the degenerate star's surface, but that's probably not enough to affect us here on Earth. Also, Sirius A is most likely tens or hundreds of millions of years away from becoming a red giant. As far as we know, there are no other Type I candidates anywhere near us.

So once again, you can breathe a sigh of relief. We appear to be safe from this kind of supernova as well.

COSMIC-RAY GUN

There is one last weapon available to both types of supernovae to consider, and it may be the most destructive yet.

Interstellar space is filled with subatomic particles—protons, neutrons, even whole helium nuclei—moving at high speeds, sometimes within a whisper of the speed of light. Called *cosmic rays* (or just CRs), they were discovered by a scientist named Viktor Hess in 1912. He lofted a balloon with a simple apparatus that detected ionizing radiation, subatomic particles capable of smacking into normal atoms and stripping them of their electrons. It was thought that most of this radiation would be near the ground (because of natural radioactive elements in the Earth), but as the balloon got higher the radiation level *increased.* That means a lot of that radiation must come from space.

What could accelerate particles to such high speeds? Why, it would take the energy of an exploding star . . . oh, right.

As mentioned before, when a star explodes, massive shock waves bounce around in the ejected material. A shock wave can dump a lot of energy into these particles, accelerating them. In the turbulent chaos of the expanding gas, a particle can get tossed around many times by shock waves, giving it a terrifying amount of velocity. When it finally escapes, it can be shot out at 99.9999 percent of the speed of light.

It's essentially a subatomic bullet, and supernovae make them by the gigaton. And it turns out that they are very dangerous indeed, because there are several ways they can hurt us here on Earth.

When CRs slam into our atmosphere, they can ionize the molecules in it and even disrupt them. Ozone, for example, is destroyed when a CR hits it. Models of nearby supernovae show that the effects from cosmic rays damaging the ozone layer are similar to those from gamma rays. Remember, anything more than about 25 light-years from a supernova is safe from its gamma rays, so we can assume the ozone will survive a cosmic-ray onslaught from such an event farther away than that.

However, when a CR hits a molecule in our atmosphere, it can create lots of dangerous high-speed secondary particles as well. These spread out like shrapnel, distributing the destruction over a larger area. These secondary particles, called *muons,* can shower down all the way to the surface of the Earth. This can be extraordinarily dangerous: muons will slam into tissue, destroying cells and DNA willy-nilly. A big enough wave of cosmic rays hitting the Earth's atmosphere could radiate muons all over the planet, killing vast numbers of plants and animals.

This type of interaction is very difficult to model. Cosmic rays are affected by magnetic fields, for example, which can alter their trajectory and speed. The galaxy has very complicated magnetic fields, and it's unknown precisely how this will affect us. The Sun's and even the Earth's magnetic fields also play into this, making it an incredibly complicated game. Still, scientists have tried to assess the situation, and because of all the uncertainties the numbers have a pretty wide range: some models show a supernova would have to be only a few light-years away to hurt us via cosmic-ray assault, while others put the distance closer to 1,000 light-years. I won't lie to you: that's not terribly reassuring, since there are plenty of such stars within that distance that can explode (as the table in the appendix indicates).

However, we can look to history for some reassurance. The sheer amount of radiation predicted by the most dire models would practically wipe out all life on Earth; muons are incredibly penetrating, so digging deep into the ground or going deep underwater to hide out doesn't help that much. Since we're here, that's pretty good evidence that the milder models are more accurate.

However, there are other effects of CRs we need to consider. As mentioned in chapter 2, when ozone is broken up by incoming cosmic rays, it can form nitrogen dioxide, which turns into nitric acid. Even a relatively mild cosmic-ray event from a supernova could increase the amount of acid rain that falls. However, if the numbers for muon events are rough, the models for acid rain from a supernova are even more poorly determined. Odds are, a supernova would still have to be pretty close to inflict this damage on us, but just *how* close is still a matter for discussion.

BLAST FROM THE PAST

Finally, let us consider one more thing. Although right now there are no potential supernovae of either type close enough to kill us, that doesn't preclude any having been too close *in the past.* The Earth is about 4.6 billion years old, and stars change their distances from each other as they orbit the galaxy like cars on a highway. Could there have been a nearby supernova sometime in the distant past that had some impact on the Earth?

Statistically, it's almost a dead certainty. Depending on the distance (the closer they are, the rarer they would be), it's possible that the Earth has had several front-row-seat views of exploding stars. One model predicts that the Earth has seen at least three within a distance of 25 light-years, close enough to severely damage our ozone layer or irradiate us with muons.

But we have more than just math to go by. We have geology.

In 2004, the scientific community received a jolt when it was announced that a team of scientists had found an anomalously high amount of the radioactive isotope ^{60}Fe (iron 60) in a sample of the seabed taken in the Pacific Ocean. The isotope is exceptionally rare on Earth, and no known terrestrial process can make it in detectable amounts.

However, this isotope *is* produced in a supernova when explosive fusion occurs in the expanding debris. It seems likely that the ^{60}Fe found in the Pacific sample was created by a supernova, and deposited when the debris swept over the Earth.

What's so very interesting about this is that ^{60}Fe has a relatively short half-life. Radioactive elements decay, producing "daughter" elements. Over time, all of the original element is gone. The half-life is the statistical time it takes for half a sample to decay, and is different for different isotopes. For ^{60}Fe, the half-life is only about 1.5 million years. By measuring how much ^{60}Fe there is compared to other elements found in the sample, the age of the sample can be determined. In this case, the ^{60}Fe sank to the bottom of the Pacific just 2.8 million years ago.

This means that a nearby supernova went off in relatively recent times, geologically speaking. Given the amount of ^{60}Fe in the sample, the supernova couldn't have been very far away either: perhaps as close as 50 light-years. Maybe closer.

In fact, the possible birthplace of the supernova has been found: a loosely knit cluster of massive stars—the kind that explode as Type II—

called the Scorpius-Centaurus Association. This grouping of stars is currently about 400 to 500 light-years away, but it was closer to Earth three million years ago—just about 100 light-years away, putting it suspiciously near the right spot for a supernova to inject ^{60}Fe onto the Earth.

Moreover, it's known that the Sun sits in a region of space called the Local Bubble: a cavity in the usual fog of gas and dust permeating the galaxy. Bubbles like this can be carved out by exploding stars; the expanding gas pushes open the cavity like a snowplow. The age of the Local Bubble, curiously enough, is less than 10 million years. The Sco-Cen Association looks pretty guilty here as well.

There are no known mass extinctions that occurred at the time the ^{60}Fe drifted onto the Earth, which is reassuring: even a supernova 50 to 100 light-years away doesn't appear to pose much of a threat.

But the statistical evidence is still interesting. Life on Earth has existed for more than 3 billion years, and multicellular life for the past 600 million or so. Was there some cosmic event sometime in that period that rocked the world?

THE CYCLE OF LIFE

But speaking of life on Earth and supernovae, there's an important point that I think we shouldn't overlook.

When the Universe began, a lot of complicated stuff happened.* At first it was too hot for even normal matter to exist; it was a soup of exotic subatomic particles. But after a short period—literally, three minutes after the Big Bang—it had cooled enough for normal matter to settle out. The early conditions were such that the only elements created at that moment were hydrogen, helium, and just a dash of lithium.

That's it. No carbon. No iron, no molybdenum, nothing but those three lightest elements. After a few hundred million years, stars formed.

---

* Really, it's true. The details aren't important here, but we'll cover them in the last chapter. Promise.

These were supermassive stars, a hundred or more times the mass of the Sun, and they were made of just these three elements; in fact they were about 75 percent hydrogen and 25 percent helium, with lithium barely even registering.

They did the usual (for today, that is) cycle of creating heavier elements out of lighter ones, all the way to iron. Then they exploded, of course, and when they did, they scattered all those heavy elements out into space. That debris slammed into nearby gas clouds, compressing them. The clouds formed the next generation of stars. These stars were different, though: they started out with some extra heavy elements in them. Some of these stars too were massive, and exploded, seeding space again with iron, carbon, calcium . . .

Eventually, the Sun was born. The Universe was already over nine billion years old at that point. Several generations of stars had polluted interstellar space with those heavy elements, so when the Sun coalesced it already had a pantry full of the periodic table. In fact, the disk from which it formed was loaded with such things as iron, silicon, and oxygen. When the planets formed from that disk, they got their share too. So the Earth is chock-full of iron, nickel, zinc, calcium, and all the rest.

But those materials didn't exist when the Universe began! It took those supermassive stars to create them. These stars were the alchemists of their day, transforming simple chemicals into more complicated ones: hydrogen became helium, became carbon, became oxygen. All the way up to iron and beyond.

When you cut your finger and a thin rivulet of blood seeps up into the slice, the red color you see is due to hemoglobin, and the key factor in that molecule is iron. *That iron was forged in the heart of a supernova.* There is enough iron created in a supernova to make well over five thousand Earths.

The calcium in your bones was most likely created in a Type I supernova, which tends to make more of that element than a Type II does. In fact, a typical Type I supernova makes enough calcium to create about 6×10^{28} gallons (that's 60 octillion gallons) of milk.

Yeah, we've got milk.

The gold in your wedding ring? Supernova. The lead in your fishing weight? Supernova. The aluminum in your foil? Well, that was probably from a red giant (they create aluminum in their cores and blow it into space in their stellar winds), but supernovae make aluminum as well.

A nearby supernova could cause destruction on an unimaginable scale . . . but without supernovae, *there would be no life in the Universe at all.* We owe our very existence to a chain of unnamed and unobservable supernovae, massive stars that died long before the Sun was more than a wisp of vapor.

It's okay to be a little scared of supernovae. But it's also okay to appreciate them. If supernovae didn't happen, who would be around to understand them?

CHAPTER 4

Cosmic Blowtorches: Gamma-Ray Bursts

T*HE BEAM CAME WITHOUT WARNING.*

There could be no warning: the wave front was moving at the velocity of light, the ultimate speed limit in the Universe. Nothing can move faster, so the wave of death brought its own announcement.

Across the Earth's southern hemisphere, people were having a normal day: shopping, working, playing, walking, hunting. When the beam reached Earth, that all changed instantly. The sky looked perfectly normal for one second, and then literally in the next it suddenly lit up, like a switch had been flipped. An intensely bright spot flashed in the sky, so bright that anyone looking at it instinctively looked away, eyes watering from the onslaught.

The new star in the sky was so fantastically bright it could outshine the full Moon, but didn't last long. It started to fade after less than half a minute, and was bearable to the eye after a few minutes. People stood in the streets, in the deserts, in the Antarctic plains, on ships at sea in the South Pacific and Indian oceans, and boggled at the incredibly bright but rapidly dimming new star in the sky.

But their amazement soon faded, and they began to go about the daily business of their lives.

Most people had already put the event out of their minds when, hours later, a flood of subatomic particles from the fading star slammed into the Earth's atmosphere. Invisibly, these particles rained down out of the sky, covering the Earth from the south pole to 30 degrees north of the equator. Australia, New Zealand, South America, essentially all of Africa and India, and half of China were blanketed with a lethal dose of radiation. It didn't matter if people were inside their houses, or outside under a clear sky: all of them were exposed.

Across two-thirds of the Earth, people started dying.

North America, Europe, and much of Asia were spared the immediate effects, but it hardly mattered. With most of the human population dying, the impact on the globe was overwhelming. And those who weren't killed outright by the burst of radiation were doomed anyway: the Earth's ozone layer disintegrated in the onslaught, dropping to half its usual strength. Ultraviolet light from the Sun was able to penetrate almost freely to the Earth's surface, killing off the base of the food chain.

The final blow was yet to come. Spawned by the wave of subatomic particles, a thick layer of smog began to form in the air, and within days the sky was a dank reddish-brown color over the whole planet. Any hardy plants that had managed to stay alive thus far suddenly found the sunlight and temperature dropping . . . which was bad enough, until the acid rain began.

And that was short-lived as well. Within weeks, the Earth's temperature had dropped enough that a new ice age was triggered. It wasn't long before the glaciers started their march from both poles.

The people who had survived the initial months of the event learned that they had witnessed the death of the supermassive star Eta Carinae, but that knowledge didn't help them. The mass extinction the star triggered would be the worst the Earth had ever seen, and when it was finally over, there were no humans left to wonder at how a single star trillions of miles away could destroy all of history in less than a minute.

COLD WAR, HOT NEWS

By the 1960s, the situation between the United States and the Soviet Union was grim. The USSR had put a base in Cuba, less than a hundred miles off the Florida coast. A failed invasion by the United States hadn't helped. Both superpowers were testing nuclear weapons on, beneath, and above the surface of the Earth. The USSR had exploded the largest thermonuclear bomb in history, equivalent to the detonation of 50 million tons of TNT.*

Needless to say, people on both sides were nervous. The end of the world by our own hand was a very real possibility.

So, in August of 1963, the United States, the United Kingdom, and the USSR signed the historic Nuclear Test Ban Treaty, limiting testing of such weapons. The very first article of the treaty states:

> Each of the Parties to this Treaty undertakes to prohibit, to prevent, and not to carry out any nuclear weapon test explosion, or any other nuclear explosion, at any place under its jurisdiction or control [. . .] in the atmosphere; beyond its limits, including outer space; or under water, including territorial waters or high seas.

This was a serious restriction. Even after more than a decade, the results of nuclear testing were often surprising. Weapons were tested not just to increase the explosive yield and improve other engineering issues, but also to see what their effects were on the environment. Just the year before the treaty was signed, in 1962, the United States had exploded a device called "Starfish Prime" 250 miles above a remote location in the Pacific Ocean. This height is essentially in space; the Earth's atmosphere is extremely tenuous that far above the surface. Starfish Prime had the relatively small yield of 1.4 megatons (that is, equivalent to 1.4 million tons of TNT), yet the effects were profound. A vast pulse of gamma rays, extremely high-energy photons of light, was created in the blast. This wave of gamma rays slammed into the Earth's atmosphere, blasting electrons off their atoms. Moving charged particles

* To this day that was the largest bomb ever exploded. Theoretically, the same design for the bomb could have been ramped up to give twice that explosive yield, but thankfully it was never tested.

create magnetic fields, and the sudden surge of rapidly moving electrons generated a huge electromagnetic pulse of energy, or EMP. This surge blew out streetlights in Hawaii, fused power lines, and overloaded TVs and radios—*all from over 900 miles away.*

Testing in space was dangerous, and the long-term effects were still not understood at the time. It became more and more clear that fallout and other effects made atmospheric and near-space nuclear tests extremely unwise. The Test Ban Treaty was hailed as a major step toward world peace.

Of course, the United States trusted the Soviet Union completely, knowing they wouldn't dream of violating the treaty . . . yeah, *sure.* While the treaty was an excellent start, no one trusted anyone else at all, and each side was very suspicious of the other. In fact, American scientists pointed out that the USSR could blow up bombs on the far side of the Moon and these would be difficult to detect. The Soviets could break the treaty and the United States would never know. What to do?

Nothing feeds engineering progress like fear. The Americans quickly found a way to check up on those scheming Soviets.

While a bomb blown up behind the Moon might be hard to detect visually, its expanding debris cloud would generate quite a bit of radioactive material in space that could be detected. One such radioactive by-product would be gamma rays. Detection technology for gamma rays was relatively new in the 1960s, but it was sufficient to sniff out any of that radiation from translunar explosions. There was one catch: gamma rays from space cannot penetrate the Earth's atmosphere, so the detectors would have to be launched on a satellite.

Besides the usual problems involved with lofting detectors into space, there was also the issue of accounting for gamma rays emitted by astronomical sources, and not from Soviet nukes. The Sun emits gamma rays, and high-energy particles from solar flares can be mistaken for them as well. A satellite might see a sudden jump in gamma rays, only to have been fooled by a solar eruption or a random particle hit.

The obvious solution was to launch gamma-ray satellites in pairs. A random particle hitting one satellite would not be seen by the other, providing a check against false detections. The data from each satellite could be compared, and if both saw an event, scientists could assume it potentially came from a noncosmic source. Other, existing satellites tracked solar flares, so those could be consulted as well.

The pairs of satellites were quickly constructed and launched. Named Vela—"watch," in Spanish—the first set was launched just days after the Test Ban Treaty was signed. They were initially crude, only able to positively detect gamma rays after taking an "exposure" of 32 seconds. But things progressed swiftly, and by 1967 the fourth pair had been launched, with a fifth—highly advanced compared to the earlier missions—ready to go.

Two scientists, Roy Olson and Ray Klebesadel, were assigned the laborious task of comparing the observations of one satellite with those of its mate. As they checked, signal after signal turned out to be negative. But in 1969, they found their first hit. The Vela 4 satellites both registered a gamma-ray event from July 2, 1967, shortly after they were initially launched. A quick look at solar flare data revealed no activity that day. Later, they found that the still-flying Vela 3 satellite pair saw the event as well.

There was one problem—whatever caused the gamma-ray event didn't look like a nuclear blast. The amount of gamma radiation and how it fades with time are very distinctive for a nuclear weapon, and the July 2 event looked completely different. There was a strong, sharp peak of emission lasting less than a second, followed by a longer, weaker pulse lasting for several more seconds.

What could this be? Unfortunately, the Vela 4 satellites couldn't tell from what direction the radiation came, so there was no way of determining the source. It may have come from behind the Moon, as feared for a nuclear test, or it may have come from some other spot in the sky entirely. Also, the event began and ended so quickly that there was no prayer of using an optical telescope to find it.

However, the Vela 5 and 6 satellites were more powerful—they were more sensitive to gamma rays, and had better time resolution. If the

July 2 event repeated, or something else like it occurred, Velas 5 and 6 had a much better shot at figuring out what was going on. Deciding that discretion was the better part of valor, the scientists waited to release the July 2 event data.

It was a good choice. Over the next few years, several more of these mysterious bursts were detected. Plus, there was an added benefit to having more satellites flying: since they were separated by thousands of miles, a crude direction could be determined for each flash. Even at the speed of light, it takes a finite amount of time for a pulse of radiation to get from one satellite to the next. That time delay, together with the known positions and separations of the satellites, could be used to triangulate on the direction of the event.

As the data built up, the scientists were astonished: the gamma-ray flares were originating from random spots in space! None appeared to come from the Sun or the Moon. It became clear that what Olson and Klebesadel were seeing was some totally unknown but extremely powerful astronomical event that no one had any previous clue about. It seemed ridiculous—how could the Universe hide such a thing from the prying eyes of astronomers?—yet there they were.

By 1973, Klebesadel and Olson had accumulated enough data to go public with the news. Together with another scientist named Ian Strong they presented the results at a meeting of astronomers in Ohio, and published a paper titled "Observations of Gamma-Ray Bursts of Cosmic Origin" in the prestigious *Astrophysical Journal*. The paper outlined the sixteen bursts they had seen up to that time (by 1979, when the Vela missions finally ended, over seventy gamma-ray bursts, or GRBs, had been detected by the satellites).

It should be noted that several other astronomers had found weird gamma-ray emissions in their detectors on various satellites as well, but couldn't be sure what they were. It took the accumulated high-quality data from the Vela satellites to be able to determine that these events were coming from deep space, or at least from outside the Earth-Moon system.

Not that the scientists had a clue what these things *actually* were.

GRBs are confusing today as well as in those early days. When Klebesadel's team released their results, the origin of GRBs was a complete mystery. Gamma rays can only be generated by high-energy events like exploding stars, solar flares, or nuclear weapons. But they had established that none were from the Sun, and none were associated with any supernovae. And they clearly weren't nuclear tests—the Vela satellites did detect several atmospheric weapons tests (from other countries), but the signals for those were unambiguous.*

What could the bursts be? To make matters more confusing, the distances to sources of GRBs were completely unknown. It was hard to imagine they were really close by (say, inside the solar system), because it didn't seem like any object or event could generate gamma rays that we wouldn't already know about. And again, the data didn't link the bursts to any observed astronomical events farther away.

As mundane explanations fell by the wayside, odder ideas were proposed. Maybe the bursts were from comets hitting the surfaces of superdense neutron stars, or maybe they were from some other equally exotic event. No one knew. But one thing most astronomers at the time agreed upon was that GRBs were not *very* far away—that is, from outside the galaxy. The farther away a source is, the brighter it must be for us to detect it. For a GRB to be outside our galaxy meant it had to generate literally unbelievable amounts of energy.

But this didn't help much. There were still too many unknowns.

There were two fundamental problems with determining the origins of GRBs: the lack of *real-time* information, and the lack of *directional* information.

The former was a significant problem. The time it took for the information from the satellites to be beamed to Earth, recorded, and then interpreted could be measured in days, or even weeks (or, in the case of

---

* One incident, however, was never clearly defined. In September 1979, what appeared to be a nuclear test well south of the Cape of Good Hope in Africa was detected, but the data were just ambiguous enough that no firm conclusion was ever made. To this day, the origin of this event is unknown.

the first one, two years). The GRBs, however, faded away in mere seconds! By the time the burst was confirmed, it was long gone. There was hope that perhaps GRBs emitted light in other wavelengths—X-rays, or optical light—and that this glow would persist long enough to be seen by other telescopes. Assuming GRBs were some sort of explosion, it would make sense that there would be an afterglow, giving astronomers time to find it. But that leads to the second problem: where to look?

The gamma-ray detectors of the time had poor eyesight: early missions simply couldn't see the direction from which the gamma rays came.

Optical light—the kind we see—has a relatively low energy. Carefully aligned lenses or mirrors inside a telescope bend or reflect the light, bringing it to a focus. This can be used to very accurately measure the position of a source of optical light. Gamma rays, however, are more like bullets zipping around. Changing their paths is much harder, and even today focusing them is beyond our technology.

What this means is that while a gamma ray can be detected and counted, getting a direction from whence it came is very difficult. Only the crudest of directions could be obtained by the Vela satellites (it wasn't much better than "somewhere over there"*). But the direction is *critical* to understanding the object. If the gamma-ray source's position is known, other telescopes can be trained at that spot on the sky to see what's what. Then any visible source seen there can be compared to known sources like galaxies or stars listed in existing catalogs. But some degree of precision is required: if the position of the burst can only be nailed down to, say, an area on the sky the same size as the full Moon, there are still thousands or even millions of objects detectable by a big optical telescope.

---

* In fact, it was worse than that. Using timing delays in signal detection actually yields two different positions for a GRB; you need at least one more satellite to distinguish between the two and determine which one is correct. It's a bit like asking, "What's the square root of 25?" Both +5 and −5 are correct. And even if you use multiple satellites, the direction to the burst is at best only a very rough estimate.

Eventually, technology started catching up to the problem. In 1991, NASA launched the Compton Gamma Ray Observatory satellite, which had GRB detectors on it. Compton's ability to get the positions of GRBs was still not great—it could only nail them down to an area on the sky the size of a quarter held at arm's length—but it was a definite improvement. Over the course of the mission, it detected over 2,700 GRBs. And while the directions were not precise, just getting that sheer number of observations was a huge advance; after enough bursts were detected, patterns began to emerge.

For one thing, that large collection of bursts allowed scientists to determine that there appeared to be two kinds of GRBs: short ones, lasting in general less than two seconds; and long ones, which lasted more than two seconds. Some bursts were even found to emit gamma rays for several minutes. As more GRBs were observed, it was found that the shorter bursts tended to give off higher-energy ("harder") gamma rays, and the longer bursts had lower-energy ("softer") gamma rays. While it wasn't understood why this might be, it was an important clue to their origins.

But the big scientific result from Compton's observations was perhaps far more important in solving the riddle: it saw GRBs spread out evenly across the entire sky. At first glance this may not seem to help, but in fact it eliminates many possibilities for their origins.

Imagine standing in a field, and insects are buzzing around. If you're in the center of the field, then you'd expect, on average, to see the same number of insects no matter what direction you look. But if you're close to the eastern edge of the field, you will see far more insects to the west (looking out across the length of the field) than to the east (looking out over the edge). The number of bugs you see in a given direction tells you something about your placement in the bug swarm (assuming the swarm is relatively random and symmetric).

So the information from Compton—that GRBs were spread randomly across the sky—instantly tells us an important fact: we are in the center of the GRB distribution *in space.*

If GRBs were inside our solar system, we'd expect to see more in one direction than another, because we are not in the center of the solar system—the Sun is. We're offset from the center by nearly a hundred million miles, and you'd expect to see that reflected in the distribution of GRBs. But there is no offset, so they are not coming from objects in our solar system.

But this also means that GRBs are not coming from sources spread around inside our Milky Way Galaxy. Since the Earth is halfway to the edge of the galaxy, GRBs in that case would be seen preferentially toward the center of the galaxy as viewed from Earth. They aren't, so they are not galactic in origin either.

That doesn't leave too many options. They could come from stars *very* near the Sun, like only a few light-years away, but not from farther stars, say, more than a few hundred light-years away, because then we'd start seeing more toward the galactic center. The other choice is that GRBs are very, *very* far away, from well outside the galaxy, millions of light-years distant.

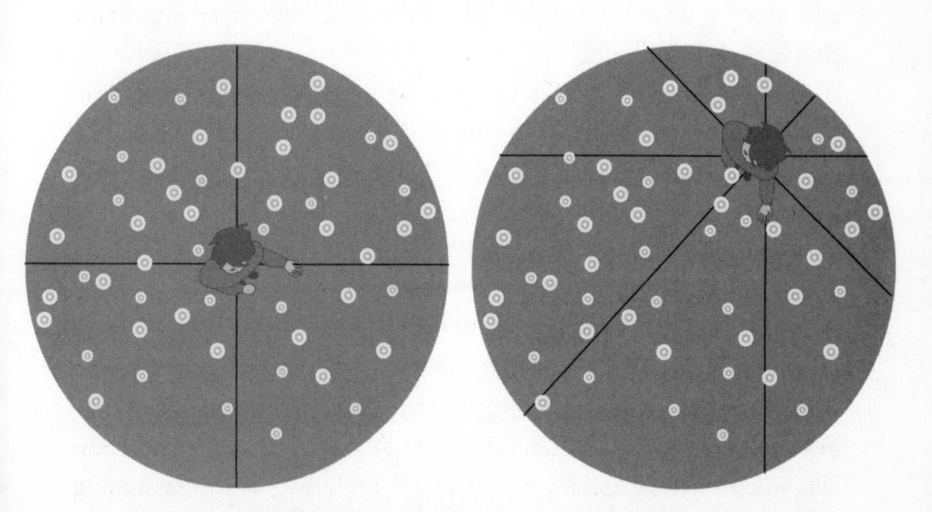

If you are in the middle of a field of fireflies *(left)*, you see equal numbers of bugs in every direction you look. But if you are off-center in the cloud of bugs *(right)*, you see more in one direction than in another. This information can be used to determine the shape of the cloud of bugs—or, more practically (to an astronomer), the distribution of GRBs in the Universe.

AURORE SIMONNET AND THE SONOMA STATE UNIVERSITY EDUCATION AND PUBLIC OUTREACH GROUP

Neither of these options is terribly palatable either. Stars shouldn't be able to make such high-energy bursts, and if they were really far away, the intrinsic energy emitted would be ridiculously high.

Still, astronomers staked their claims on either side of this issue, publishing papers furiously and arguing—sometimes also furiously—over it. They even staged a famous debate about it between two accomplished scientists who took different sides of the debate: one defending the idea that they were from nearby stars, the other saying they were coming from the distant reaches of the Universe. But even by the time the debate was held, preparations were under way to get the real answers.

THE VIEW FROM AFAR

The joint Dutch-Italian satellite BeppoSAX was launched in 1996. While it was not designed specifically to hunt for GRBs, it had that capability. More important, it had on board a revolution waiting to happen: detectors that could actually get a good direction for incoming X-rays (which, like their higher-energy brethren, gamma rays, are difficult to pin down). It also had a wide field of view, which increased the odds of detecting a randomly placed burst, even if the position was not well known at first.

In February 1997, a long GRB was detected by the BeppoSAX monitor. It also happened to lie within the field of view of the X-ray detectors. Observations were made, and then repeated a few days later. Breakthrough! The results were clear—a bright source of X-rays had faded considerably in the interval. Astronomers knew that must be from the fading afterglow of the burst. And better yet, the X-ray detectors were able to get a reasonably good position for the burst, now called GRB 970228 (for the gamma-ray burst seen in 1997 on February 28).

Within a month, the Hubble Space Telescope was pointed at the location of the GRB and the breakthrough got more momentum: a fading glow in *visible* light was detected, and it appeared to be right next to a dim, distant galaxy. This was too close to be a coincidence.

Then, finally, the clincher. In May of that same year, the mammoth ten-meter Keck telescope in Hawaii obtained spectra* of a GRB afterglow. This allowed astronomers to determine an accurate distance to GRB 970228, and they were astonished to see that it was located a numbing *nine billion light-years away.* That's more than halfway across the Universe!

Finally, after thirty years, thousands of burst observations, and countless arguments, a major question was answered: bursts were not only far away, they were *very* far away. After this, no one doubted the vast distances to gamma-ray bursts. They were coming from well outside our Milky Way, and in fact close to the visible edge of the Universe.

But that left one problem, vast in its own right: what event could *possibly* generate such incredible energies?

KABOOM!

No matter how you slice it, GRBs are, for a short time, the most luminous objects in the Universe, the best bangs since the Big One.

This is no small problem. Imagine a source of light in space: the light it emits will expand as a sphere with the source at the center. As the sphere grows, the light gets spread out, and will appear dimmer to an observer (that's why lights get fainter with distance). When the distance to the object doubles, the area over which the light spreads out goes up by four times,† so the brightness will dim by four times. If you increase your distance to 10 times farther away, the light will be only one-hundredth (1 percent) as bright, and so on. The brightness of an

---

* A spectrum is what you get when you run light through a prism or a finely etched grating. The light is separated into its individual colors, like a rainbow. When measured very carefully, a wealth of information can be obtained about the source of the light, including its temperature, chemical composition, and for some objects, like galaxies and GRBs, even their distance.

† Stretch your brain back to dim memories of high school math: the area of a sphere = $4\pi r^2$.

object therefore decreases *very* rapidly with distance. This presented a serious problem for GRB researchers: from a distance of billions of light-years, the explosion that formed the GRB must be *huge* to be able to be detected at all from Earth. When the numbers were crunched, it didn't make sense. Even *converting an entire star into energy* using Einstein's $E = mc^2$ (see chapter 2) wouldn't provide enough energy to fuel the burst, and that is literally the most energy you can get from a star (ignoring the inconvenient fact that there's no known way to convert an entire star into energy, and certainly not in the span of a few seconds).

But there was still an out. What if the blast *wasn't* symmetric, expanding equally in all directions? What if it was *beamed*?

If you take a small lightbulb and turn it on, it emits light in all directions, and its apparent brightness fades rapidly with distance. But if you put it in a flashlight, which collects its light and focuses it into a beam, the light appears brighter from farther away.

Astronomers could almost taste the answer to this piece of the GRB puzzle. Instead of a nearly impossibly energetic blast at a colossal distance expanding spherically and fading rapidly, *maybe the explosion was less energetic, but focused into beams.* Beaming would mean only a tiny fraction of the energy would be needed compared to a spherical blast.

The energy of the detonation would still have to be frighteningly huge to be seen clear across the Universe, but not impossibly so. In fact, the energy involved would be similar to that of a supernova. This gave astronomers hope that they could find the Holy Grail of GRB science: the engine that drove this phenomenon.

And of all the objects in the cosmic zoo that astronomers knew of, only one could possibly generate those kinds of forces.

THE GRAVITY OF THE SITUATION

Black holes are notorious for sucking down matter and energy, not spewing them out, so it might seem paradoxical that they could be at the heart of gamma-ray bursts, the brightest objects in the Universe.

But the key to this is gravity. And the key to *that* is how black holes form, so let's take a step back (a good idea when dealing with black holes) and take a look at this singular event.

In chapter 3, we saw that massive stars explode when they run out of fuel to fuse in their core. The incredibly strong gravity of the core makes it collapse, which sets off a series of events that blows up the star. That description focused mostly on what happens to the outer layers in a supernova, but not what happens to the core itself. But it's there that the power of the GRB lies.

As the iron core of the incipient supernova collapses, the electrons are rammed into the protons, making neutrons (and emitting neutrinos, the major trigger of the supernova explosion). In a flash the entire core of the star becomes a sea of neutrons with almost no normal matter left. What was once a ball of iron thousands of miles across is now an ultradense neutron star, perhaps ten miles across. It has a mass equal to the Sun, but a density magnified beyond belief: a spoonful of neutron star matter would weigh *a billion tons*! That is somewhat more than the combined mass of every single car in the United States—imagine 200 million cars crushed down to the size of a sugarcube and you'll start to get an idea of how extreme neutron star matter is.

The neutron star's incredible mass is supported by a weird quantum mechanical effect called *degeneracy* (see chapter 3). It is similar to electrostatic repulsion—the idea that like charges repel—but instead it's a property of certain subatomic particles where they resist being squeezed too tightly together. Degeneracy will occur if you try to pack too many electrons together, but it also affects neutral particles like neutrons. It's an astonishingly strong force, and is able to keep the vast bulk of the core from collapsing further. The collapsing core of the star slams to a halt, and a neutron star is born . . .

. . . most of the time. It turns out that if the mass of the collapsing stellar core exceeds about 2.8 times the Sun's mass, even neutron degeneracy cannot hold it up. The core's gravity is too strong, and the core collapse continues. This time there is no force in the Universe strong enough to stop it.

What happens next is so bizarre that it stretches the human mind to its limit to understand. As an object gets smaller, but retains its mass, its gravity gets stronger. As an easy example, if you were to somehow shrink the Earth to half its current diameter while still keeping its mass, the gravity you feel (and therefore your weight) would increase. The smaller the Earth gets, the stronger its gravity.

If you wanted to launch a rocket to the Moon from this newly shrunken globe, you'd have to give it a lot more power to overcome the Earth's gravity. If you shrank the Earth more, the rocket would need even more power, and so on. Eventually, the Earth would shrink so much that its gravity would be literally impossible to overcome.

You might think you just need to add more thrust to the rocket, but when matter gets this dense, Einstein has something to say about the situation. He postulated that gravity is really just a manifestation of bent space. What you feel as a force downward, toward the center of the Earth, is actually a bending of space, like the way the surface of a mattress would bend if you plunked a bowling ball down on it. Roll a marble across the bed, and the path of the marble bends, just the same way an asteroid's path bends because of gravity when it passes near the Earth.

This is more than just a model, more than mere speculation. Its consequences are very real: if too much matter is packed into too small a volume, the bending of space can become so severe that it literally becomes an infinitely deep pit. You can fall in, but you can never climb back out.

An object like this is like a hole in space. Nothing can escape it, not even light. Since it cannot emit light, this hole would be black. What would *you* name such a thing?

And so it goes in the core of the exploding star. If the core is too massive to form a stable neutron star, it collapses. *All* the way down. It shrinks to a mathematical point, space gets bent to the breaking point, and a black hole is born.

The gravity of the hole is intense. Any matter close by will be drawn inexorably into it. But there's a hitch. Stars spin, and so do their cores. As the core collapses on its way to forming a black hole, that rotation increases, the same way an ice skater can increase her spin by drawing her arms in. Once the black hole is created, it will be spinning very rapidly, and any matter falling in will *also* revolve around it, like water going down a drain. The closer to the black hole it gets, the faster that matter will swirl around it.

So matter falling into a black hole doesn't just fall straight in—plonk!—and disappear; it *spirals* in. The matter just outside the black hole begins to pile up, and it forms a flattened disk called an *accretion disk* (accretion is the process of accumulating matter). This will happen for any star that is spinning before it collapses, but models have shown that GRB progenitors may be spinning even faster than normal. These rapid rotators form an accretion disk much more readily than a slowly rotating star. And once the disk forms, the ferocious gravity of the black hole will get the inner part of the disk moving very close to the speed of light, and even matter farther out from The Edge will still be moving incredibly rapidly.

When a black hole forms, spin and gravity are not the only things to get amplified. Stars also have magnetic fields, like giant bar magnets (see chapter 2). Just as gravity increases as the star shrinks, so does the magnetic field. A typical star may have a magnetic field not much stronger than the Earth's: just enough to make a needle in a compass move. But if you take a star a few million miles across and squeeze it into a ball just a few miles across, the magnetism increases hugely as well, getting billions and even *trillions* of times stronger.

Any matter trying to fall into the black hole is therefore under the influence of a witches' brew of forces. Gravity tries to draw it in, but its angular momentum counteracts that, forming the disk. The magnetic fields also get twisted up like a tornado as the matter spins around the disk. And on top of it all, there's just plain old heat, created, oddly, by something familiar amid all these exotic forces: friction. As the matter in the disk swirls madly around under the force of the black hole's gravity, the particles in it slam together at incredibly high speeds, which generates immense quantities of friction. This heats the disk to millions of degrees.

The sheer heat tends to drive particles away from the black hole. If a particle tries to move outward in the plane of the disk, it slams into other particles and cannot escape. But if it goes *up*, out, it's free to travel—there's less material in that direction. Moreover, the monstrously amplified magnetic fields can also accelerate the particles up and out. The heat and magnetism combine to focus a pair of tight beams, like two ultra-mega-superflashlights glued together at their bases. These twin beams shoot out from directly above and below the black hole, firing outward, away from the black hole in directions perpendicular to the disk.

What happens next is a vision of hell so apocalyptic that it's difficult to exaggerate. Moments after the black hole is created and the disk forms around it, all that energy—a billion *billion* times the Sun's output—is focused into twin beams of unmitigated fury. So much energy is packed so tightly into the beams that they blast outward in opposite directions, eating their way through the star at the speed of light. Within seconds, the beams have chewed their way out to the surface and are free. Any matter in their way is torn apart, heated to billions of degrees, rendered into its constituent subatomic particles, and accelerated to within a hairbreadth of the speed of light. Ironically, by the time they punch their way through the star, perhaps only a few hundred Earth-masses of matter are in the beam, which is huge on a human scale, but tiny on a cosmic one. But that also is a key to their power: since the total amount of matter in the beams is relatively small, it can be accelerated to incredible speeds.

Clouds of gas still surround the doomed star, echoes of past eruptions before the final explosion. The beams of energy and matter slam into this material, creating vast shock waves, sonic booms in the material, but on a mind-numbing scale.

There are also shock waves generated inside the jet itself as parts of it move faster than others. When these collide, the awesome energy of the jet churns up the matter inside them, creating unimaginable turbulence, which in turn adds greatly to the energy emitted. The ensuing conflagration emits gamma rays, huge amounts of them, as the magnetic fields and raw energy of the beams bombard the matter.

When a *very* massive star's core collapses, twin beams of matter and energy can be focused by the incredible forces in the star's center. The beams may last only a few seconds, but contain as much energy as the Sun will emit in its entire lifetime, or more.

DANA BERRY, SKYWORKS DIGITAL INC.

A gamma-ray burst is born.

The beams continue on. Behind them, the rest of the star finishes its collapse, forming what would otherwise be a normal supernova. Before the discovery of GRBs, a supernova was considered the most violent, the most energetic single event in the Universe. But a decent GRB can dwarf the energy of even a supernova. Because of this, astronomers coined a new word to describe the event: *hypernova*.

Once the beams pass through the gas, they continue on, leaving behind superheated matter that begins to cool. As it does, it emits light for some time after the beams have moved on. This is the source of the afterglow sought so dearly by scientists on Earth. The matter can get extremely bright—one GRB in 2008 was nearly 8 billion light-years away, but was visible to the naked eye! But the afterglows fade rapidly, dropping in brightness by factors of thousands in just a few minutes.

That's why the optical afterglow was initially so difficult to detect. Even the titanic energy of a GRB is mitigated by raw distance.

But we now know that GRBs are created in a hypernova, when a massive star explodes . . . and we see massive stars in our own galaxy. Sure, all the GRBs we have ever seen have been at terribly remote distances, billions of light-years removed from Earth.

But what happens if one goes off that's *not* far away? What if a nearby star becomes a GRB?

BEAM ME OUT

An object that finds itself in the path of the beams of a nearby GRB will have bad things happen to it.

Very bad things.

But before I scare the pants off you, remember that if you are far enough away they are no danger at all. The only reason we can see GRBs at all is *because* we are in the path of the beams: since all the light of the GRB is focused into those beams, if they miss us we don't see anything. So if they are far enough away you just see a faint blip, and it's gone. But if you're too close . . .

The effects from a GRB are very similar to those of a supernova, which isn't surprising. They are related phenomena, with GRBs being produced in supernovae, and they both emit huge amounts of energy in the form of gamma rays, X-rays, and optical light.

Where they differ is how well they sow their destruction over different distances. With a supernova, which emits radiation and matter in all directions, the effects die down rapidly with distance. As we saw in chapter 3, they appear to be mostly harmless from a distance of more than 25 to 50 light-years or so.

But GRBs are beamed. Their luminosity does not decrease as rapidly with distance, and this makes them dangerous from farther away. *Much* farther away.

Every GRB is different, making prognostication difficult. But enough have been observed to do a little averaging and get the effects from a typical GRB, whatever "typical" means when you're dealing with Armageddon focused into a death ray.

Let's set the scene.

Why fool around? Let's say a GRB went off *really* close: 100 light-years away. Even from that very short distance, the beam of a GRB would be huge, 50 trillion miles across. This means that the whole Earth, the whole *solar system,* would be engulfed in the beam's maw like a sand flea in a tsunami.

GRBs, mercifully, are relatively short-lived, so the beam would impact us for anywhere from less than a second to a few minutes. The average burst lasts for about ten seconds.

This is short compared with the rotation of the Earth, so only one hemisphere would get slammed by the beam. The other hemisphere would be relatively unaffected . . . for a while, at least. The effects would be worst for locations directly under the GRB (where the burst would appear to be straight up, at the sky's zenith), and minimized where the burst was on the horizon. Still, as we'll see, no place on Earth would be entirely safe.

The raw energy that would be dumped onto the Earth is staggering, well beyond the sweatiest of cold war nightmares: it would be like blowing up *a one-megaton nuclear bomb over every square mile of the planet facing the GRB.* It's (probably) not enough to boil the oceans or strip away the Earth's atmosphere, but the devastation would be beyond comprehension.

Mind you, this is all from an object that is *600 trillion miles away.*

Anyone looking up at the sky at the moment of the burst might be blinded, although it would probably take several seconds to reach peak optical brightness, enough time to flinch and look away. Not that that would help much.

Those caught outside at that moment would be in a lot of trouble. If the heat didn't roast them—and it would—the huge influx of ultraviolet radiation would instantly give them a lethal sunburn. The ozone layer would be destroyed literally in a flash, allowing all the UV from both the GRB and the Sun down to the Earth's surface unimpeded. This would sterilize the surface of the Earth and even the oceans down to a depth of several yards.

And that's just from the UV and the heat. It seems cruel to even *mention* the far, far worse effects of gamma rays and X-rays.

Instead, let's take a step back. GRBs are incredibly rare phenomena. Although they probably happen several times a day somewhere in the Universe, it's a really big Universe. The odds of one happening 100 light-years away are currently zero. Zip. Nada. There are no stars close to us that are *anywhere near* the capability of becoming a burst. The nearest supernova candidate is farther away than that, and GRBs are far rarer than supernovae.

Feel better? Good. So let's try to be more realistic. What *is* the nearest GRB candidate?

In the southern sky is a star that looks unremarkable to the naked eye. Called Eta Carinae,* or just Eta for short, it's a faint star in a crowded field of brighter ones. However, its faintness belies its fury. It's actually about 7,500 light-years away, and is in fact the most distant star that can be seen with the unaided eye.

The star itself† is a monster: it may have 100 or more times the mass of the Sun, and it emits 5 million times the energy of the Sun—in one second, it gives off as much light as the Sun does in *two months.* Eta suffers periodic spasms, blowing off huge amounts of matter. In 1843, it underwent such a violent episode that it became the second brightest star in the sky, even at its vast distance. It expelled huge quantities of matter, more than 10 times the mass of the Sun, at over a million miles per hour. Today, we see the aftereffects of that explosion as two huge lobes of expanding matter, each looking like the blast from a cosmic

---

* Pronounced "ATE a CARE in Ay," for those practicing at home.

† Actually, Eta may be a binary star, two stars orbiting each other. There is so much glare and interference from all the material surrounding the star that astronomers still aren't 100 percent sure.

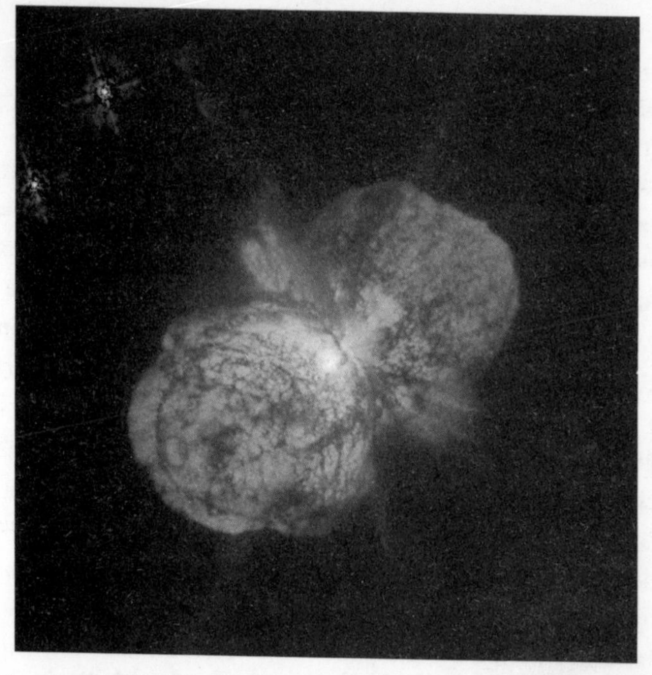

Eta Carinae is the Milky Way's scariest star. It may be a binary, with one star 100 times more massive than the Sun. In 1843, it underwent a titanic convulsion, almost as powerful as a supernova, that ejected the two lobes of matter on either side. When Eta finally blows, it may explode as a hypernova and a GRB.

JON MORSE (UNIVERSITY OF COLORADO), KRIS DAVIDSON (UNIVERSITY OF MINNESOTA), AND NASA/ESA

cannon. The energy of the event was almost as powerful as a supernova itself.

Eta has all the markings of a GRB in the making. While it will certainly explode as a supernova, it isn't known if it will be a hypernova-type GRB or not. It should also be noted that if it does explode and becomes a GRB, the orientation of the system is such that the beam will almost certainly miss the Earth. We can tell this from the geometry of the gas ejected in the 1843 paroxysm: the lobes of expanding gas are tilted with respect to us by about 45 degrees, and any GRB beaming would be along that axis. To make this more clear: we are in no danger from a GRB, Eta or otherwise, in the near or even mid-term future.

But still, it's fun to speculate. What if Eta were aimed at us, and it went hypernova? What would happen?

Again, bad things. While it wouldn't get anywhere near as bright as the Sun, it would certainly be as bright or possibly even ten times brighter than the full Moon. While this is bright enough to make you squint, it would only last a few seconds or minutes, and so would probably do no long-term damage to any flora's or fauna's life cycles.

The levels of influx of ultraviolet light would be intense, but brief. People outside would be mildly sunburned, but in all likelihood there would not be a statistically measurable increase in skin cancer rates down the line.

But the situation is very different when you look at gamma rays and X-rays. These would be absorbed by the Earth's atmosphere, and the effects would be far worse than for a nearby supernova.

The most immediate effect would be a strong electromagnetic pulse, far stronger than the one experienced in Hawaii during the Starfish Prime nuclear test. In this case, the EMP would immediately wipe out every unshielded electronic device on the hemisphere of the Earth facing the burst. Computers, phones, airplanes, cars, anything with electronic circuitry would be fried. This also includes power grids; a huge current would be induced in the transmission lines, overloading them. People would be without power, and without any means of long-distance communication (satellites would have all been fried by the gamma rays anyway). This would be more than an inconvenience, as it means hospitals, fire stations, and other emergency personnel would be without power as well.

But as you'll see in a moment, we may not have much use for emergency services . . .

The effects on the Earth's atmosphere would be severe. This situation has been extensively studied by scientists. Using the models described in chapter 3, and assuming a GRB at Eta's distance, they have determined the effects. They are not pretty.

The ozone layer would take a huge hit. The gamma rays from the

GRB would blast apart ozone molecules wholesale. Globally, the ozone layer would be reduced by an average of 35 percent, with smaller localized regions being depleted by more than 50 percent. This all by itself is incredibly damaging—mind you, the ozone troubles we have currently are due to a relatively slight dip of just 3 percent or so.

The effects of this are very long-lasting, and could persist for years—even five years later there could be as much as a 10 percent ozone depletion. During that time, ultraviolet light from the Sun would be more intense on the Earth's surface. Microorganisms that form the base of the food chain are highly susceptible to UV radiation, and would be killed in vast numbers, leading to a possible extinction-level event that would work its way up that chain.

To make matters worse, the amount of reddish-brown nitrogen dioxide formed (see chapters 2 and 3) in an Eta Carinae GRB event would actually reduce the amount of sunlight reaching the Earth by a significant amount.

The exact effect of this is hard to determine, but it seems likely that even a few percent drop in sunlight over the entire Earth (the nitrogen dioxide would spread all over the atmosphere) would cool the Earth considerably, and could conceivably start an ice age.

On top of this, enough nitric acid would be generated in the chemical mix that there would be acid rain, which would also potentially have devastating environmental effects.

Then there's the issue of subatomic particles (cosmic rays) from the burst. It's unclear just how damaging these would be from a GRB. But, as discussed in chapters 2 and 3, high-energy particles may have all sorts of effects on the Earth. A GRB from 7,500 light-years away would inject a vast number of subatomic particles into our atmosphere, and they would be moving at just a shade under the speed of light. Within hours of the initial burst they would slam into the air, creating a shower of muons. We see muons coming from the sky all the time, but in small amounts. From a nearby GRB, the number of muons generated would be huge. One team of astronomers calculated that as many as

300 billion per square inch could hit the Earth's surface all over the hemisphere facing the blast.* If that sounds like a lot, well, it is. These particles would cascade down from the sky and be absorbed by anything out in the open. Given how well human flesh can absorb muons, the astronomers who did the calculation found that the energy absorbed by an unprotected human would be ten times the lethal dose. Hiding won't help much; muons can penetrate water to depths of more than a mile and also go right into rock down to depths of half a mile! This would therefore affect nearly all life on Earth.

So in reality, ozone depletion wouldn't be that big a deal. By the time that really became a problem, most of the animals and plants on Earth would be long dead anyway.

That is the nightmare scenario depicted at the beginning of this chapter. However, before you panic, remember: Eta Carinae is almost certainly pointing in the wrong direction. But while we're on the topic, there *is* another possible GRB progenitor to consider. Called WR 104, it's coincidentally about the same distance from us as Eta. It's a binary star, and one of the stars is a bloated, massive beast near the end of its life. It *may* blow up as a GRB, and it *may* be pointed more or less at us, but those are both pretty iffy. The odds are that we're safe from this monster as well, but it's worth mentioning.

THIS IS NOW, THAT WAS THEN

So we seem to be pretty safe at the moment, which is a good thing. The odds of a nearby GRB at any given time are extremely low . . . but the Earth is old. Is it possible that we got zapped by a GRB in the past?

Statistically speaking, it's actually quite probable that the Earth was hit by a relatively close GRB beam at some point in the past. While supernovae are common enough, they have to be close to hurt us. GRBs are

---

* These findings are still controversial. This is a new field of science, and the models are somewhat shaky. Still, if you take away anything, just remember that a nearby gamma-ray burst is *bad*.

far rarer, but are damaging at much greater distances. Some studies have shown that one should be near enough to do some ecological damage to the Earth every few hundred million years or so.

It turns out that there may even be evidence for one such event in the Earth's past. The end of the dinosaurs may be the most famous mass extinction event in history, but it was not the largest. The Ordovician era ended about 440 million years ago, when as much as half of all genera of life on Earth were wiped out. It happened rapidly, and appears to have had two separate extinction events separated by perhaps a million years. The cause has mystified scientists for many years.

Could a GRB have pulled the trigger on this extinction event? There are many tantalizing clues. In a GRB event, you'd expect incoming UV radiation would more profoundly affect animals and plants that lived near the surface of the oceans than it would affect deep-sea creatures, and there is evidence in the fossil record that that is what happened. Trilobites, those curious crablike animals that dominated the oceans of the time, had a larval stage. It appears that at the time of the extinction event, larvae that lived near the surface of the water were more affected than those that lived in deeper water, indicating that whatever caused the sudden die-off may have come from above, from the sky. Moreover, animals that had longer larval periods in their life cycle were also more likely to go extinct than those with shorter larval stages. These are both consistent with a sudden increase in UV radiation that could affect shallow-water regions, but not deep-water. Animals with longer larval stages would absorb more dangerous UV radiation, which would preferentially kill them off.

Interestingly, such trends are *not* seen in other mass extinctions, indicating that the Ordovician extinction had an unusual cause. GRBs are many things, but "unusual" would be high on that list.

The second Ordovician extinction event has been associated with rapid cooling of the Earth, followed by glaciation. This is also consistent with the effects of a nearby GRB; the cosmic-ray shower and subsequent increase in atmospheric nitrogen dioxide would contribute to a possible global cooling. In fact, some researchers have found that at this time on Earth, a global glaciation could not have occurred without some sort of "forcing event"—that is, some outside mechanism to kick-start it. Perhaps that force came from a GRB.

This evidence is interesting, perhaps even persuasive, but it is not conclusive. More research, as usual, is needed. But it does give one pause to think that an event that occurred thousands of years earlier and trillions of miles away could so profoundly affect life on Earth.

BEAMING WITH CONFIDENCE

Are GRBs worth worrying about?

One answer is no, because if one goes off there's nothing we can do about it. And since gamma rays travel at the speed of light—they *are* light—we will get literally no warning if one is headed our way. So why worry?

On the other hand, it's quite possible that there is nothing to worry about anyway.

Almost every gamma-ray burst ever seen has come from an incredibly distant galaxy. But in astronomy, distance is the same thing as time: the farther away you look, the farther back in time you are seeing. When we see a GRB explode in a galaxy nine billion light-years away, we're seeing that galaxy as it was nine billion years ago. GRBs were common in the past, and became less frequent as the Universe aged.

This is significant because galaxies change over time. Early in their lives, they had fewer heavy elements like calcium, iron, and oxygen in them; these elements are created and distributed into the galaxies by supernovae, and that takes time. It turns out that it's easier for stars with fewer heavy elements to turn into GRBs when they die. Since most massive stars currently being formed have lots of heavy elements in them, thanks to previous generations of supernovae, they are *less* likely to go GRB.

Furthermore, stars that explode as GRBs need to be rotating rapidly before they collapse, or else the accretion disk that feeds the beams may not form. It turns out that stars with higher abundances of heavy elements tend not to spin so quickly. It's not because the elements are more massive, however! Heavier elements are better than lighter ones at absorbing the light coming up from the star's interior. This makes a star with lots of heavy elements in its gas hotter and brighter than a star with fewer heavy elements in it. Because of this, particles on the surface of the star are more easily blown away in a stellar wind—the equivalent of the solar wind, but from a star other than the Sun.

As the particles leave the star, they are swept up by the rotating magnetic field of the star. This acts like a parachute, slowing the star's spin in turn: imagine holding a plastic bag open and spinning around; as the bag catches the air, your spinning would slow because of the drag. The same thing happens to stars; their spin slows over time as their magnetic field drags through the stellar wind. In fact, this is why the Sun rotates only once a month. It probably spun much faster when it was young, but over billions of years the solar wind dragging through the magnetic field has slowed its rotation.

So stars that have more heavy elements have a stronger stellar wind, and tend to spin more slowly. The converse—stars with fewer heavier elements tend to rotate more rapidly—means that stars that were born earlier in the life of the Universe will make more GRBs than stars born more recently. The upshot of all this is that *GRBs from hypernovae—from massive stars exploding—will be more rare today than they were in the distant past.*

In other words, you really don't have to worry too much about *them.*

SHORT, BUT NOT SWEET

So are we safe from this form of destruction, sitting comfortably in our twelve-billion-year-old galaxy with its heavy elements and slowly spinning massive stars?

Maybe. But maybe not. If you recall, there appear to be two different kinds of GRBs, ones that last longer than two seconds, and ones that are shorter. The kind generated in the collapse of a massive star's core is the long kind of GRB. But what of the short ones?

Two NASA satellites were critical for understanding the short bursts. The High-Energy Transient Explorer-2 (HETE-2) and Swift missions detected dozens of short GRBs. Using these observations, astronomers were able to craft the idea that a short GRB can occur when two dense neutron stars merge. A neutron star forms when the core of a star going supernova is not quite massive enough to form a black hole. In many cases, massive stars form in pairs, with the two stars orbiting one another, and many such high-mass star pairs are seen in our galaxy. Over time, the more massive star will explode, leaving behind a neutron star. Some time later, the other star explodes, *also* leaving behind a neutron star.

Through many forces, over billions of years, the orbits of the two stars will shrink. The two ultradense objects spiral closer and closer together . . . and then, finally, they will get so close that they literally merge. Their combined mass may be enough to form a black hole, and if enough matter is left over it will form an accretion disk around the hole. At this point, events are similar to what happens in the core of the massive star when it explodes: the accretion disk, tremendous magnetic fields, and powerful gravity of the black hole focus twin beams that explode outward.

Models of these events indicate that the burst of gamma rays would be much shorter in duration than the massive star type of GRB, and would produce higher-energy gamma rays. Both of these predictions fit the observations. There are other models that also fit the observations (such as a black hole–neutron star binary, with similar results), but this is the leading theory.

One major difference between the merging neutron star GRBs and the massive star hypernova GRBs is the time it takes before one can go off: while in modern times we expect to see few if any massive star GRBs, we expect to see plenty of neutron star mergers. It takes billions of years for the orbits of the two neutron stars to decay and cause the stars to merge, and so they should be able to occur today. This may very well be true, but in raw numbers they are less common than their more massive counterparts. This may be due to their uncommon origin—there are plenty more single massive stars that can explode than there are *binary* massive stars—so it's difficult to get a handle on how many

Two neutron stars finally succumb to their mutual gravity after billions of years of orbiting each other. Torn apart, they merge and collapse into a black hole, which announces its birth with a GRB.

DANA BERRY, SKYWORKS DIGITAL INC.

potential short GRBs there are in our galaxy. There are many neutron star binaries known, all of which could become short, hard GRBs . . . in a few more billion years. None are known that could go off in a century or millennium, or even in the next million years. But unlike massive stars, which are incredibly bright and obvious, binary neutron stars give off very little light and are difficult to detect.

It's unlikely in the extreme that there are any close enough to do us any harm. But it's not possible to entirely rule them out either.

THE FUTURE IS BRIGHT

What we need, as always, are more observations. As the biggest explosions we know of—and probably the biggest explosions the Universe can make—GRBs are of great scientific interest. They tell us so much about how matter and energy act at the extreme limits of physics, how black holes are born and behave, and also about the environment around them. There's still a lot we don't understand about GRBs, of course. We've come a long way since the Vela satellites; in 2004, NASA launched the Swift satellite—so critical in understanding the origin of the short, hard bursts—which has observed hundreds of GRBs, including the most distant one ever seen at 12.8 billion light-years away. Swift's observations have allowed keen insight into both long and short bursts, adding much-needed data to the theoretical models.

As we learn more about GRBs, we'll be better able to assess the danger from them, including how they may have affected life on Earth in the past. While there's probably nothing we can do if one goes off—and it's incredibly unlikely it will happen at all—it's always better to have a good handle on the situation.

So should you worry? I'm asked this all the time, and I have a simple answer: I have known people who've been killed in all sorts of unlikely ways, including car crashes and one who was hit by lightning. How many people do you know who have been killed by a gamma-ray burst?

CHAPTER 5

The Bottomless Pits of Black Holes

EVER VIGILANT, IT'S AN AMATEUR ASTRONOMER WHO *catches the first whiff of trouble.*

He had hopes of doing some imaging of Uranus using his automated telescope, but the computer consistently points the telescope in the wrong direction. After going to manual, he eventually finds the giant planet several arc minutes from its calculated position. Puzzled, he calls a friend who quickly confirms that he has the same problem. An Internet search on a few astronomy bulletin boards reveals many such events from astronomers all over the world.

As days go by, things get worse. Jupiter seems to be off-kilter as well. Saturn, however, located on the other side of the Sun, appears unaffected. Rumors start to spread.

Then the situation gets really weird. The Solar and Heliospheric Observatory, parked in an orbit where the Earth's gravity and the Sun's gravity are in balance, starts to drift. Engineers are puzzled, but soon have other problems with which they must contend. Now Mars is in the wrong place. NASA has a probe on its way to the Red Planet; will it miss? But soon that point becomes moot, as the spacecraft is drifting too. After a few days it's clear the probe is lost . . . and that the probe is the least of our worries.

Solar astronomers detect that the Sun's position is off as well. That doesn't make any *sense. What could move an entire star . . . ? But they quickly realize the trouble is not with the Sun, but with the* Earth. *Like the other planets, the Earth is no longer circling the Sun as usual, but is moving off its prescribed orbit.*

Panic spreads. Scientists come to the obvious conclusion: some massive object is approaching the Earth, and its gravity is pulling us off course. They use the data on the other planets' motions to determine where this object must be, but find nothing at that location of the sky.

Ironically, seeing nothing confirms their worst fears: it's a black hole. Backtracking its position reveals it's headed almost straight at us at the incredible speed of 500 miles per second. Astronomers calculate its mass as a terrifying ten times that of the Sun's—easily enough to spell doom for us on Earth. The gravitational effects are subtle at first, but accelerate.

Just a few weeks after the first trouble began—and its position still 300 million miles away—the black hole's gravity as felt on Earth is equal to that of the Sun. Earth no longer orbits one star: it is enthralled by two: one living, one dead. Within a few more days, the black hole's influence is far stronger than the Sun's. Grasping the Earth with invisible fingers, it tears us away from the Sun, bringing us closer to the collapsed star.

As we approach, the gravitational tides from the black hole begin to stretch the Earth. Tides from the Moon cause the oceans to ebb and flow, but the black hole has 200 million times the mass of the Moon. Even from millions of miles away, the tidal force is causing enormous floods, gigantic earthquakes, tsunamis.

The coup de grâce quickly arrives. When the black hole reaches a distance of just seven million miles, the force of its gravity as felt by objects on the Earth's surface is equal to the gravity of the Earth itself. The few survivors of the past few days' events suddenly find themselves weightless as they are pulled both up and down with equal force.

Within minutes, as the black hole draws ever closer, the force upward dominates. A rising hurricane of air now blows weightless people up, along with rocks, cars, the oceans . . .

An hour later, it's all over. The immense gravity of the dead star rips the Earth to pieces, shredding it into vapor. The material that once constituted our home world falls toward the voracious maw of the hole, swirling around it ever faster, forming a disk of million-degree plasma before taking the final plunge.

Without a hiccup, without a stumble, the black hole sails on, down and out of the solar system, leaving behind chaos, scattered planets, and death.

THE HOLE TRUTH

What is it about black holes? The mind-boggling physics, the sheer destructive power, the weird way they twist our notions of reality, space, and time?

Maybe they fascinate us simply because they're *cool.*

Born in the hellish heart of a supernova, announcing their presence with twin beams of unstoppable fury, and devouring (almost) all that is in their path, black holes are firmly fixed in the public's mind. Movies, television shows, books, countless articles, and endless discussion have revolved around them. Yet with all this excitement and interest, most people really have only a vague idea of just what black holes are, and what they can and cannot do.

But never forget, they're *dangerous.* There are many ways a black hole can kill you. Some are simple, and some are truly bizarre. Unless you're looking for trouble, they're all unlikely in the extreme, but if you want rampant destruction on a large scale, then a black hole is a good place to start.

I'VE FALLEN AND I CAN'T GET UP

As pointed out in chapter 4, a black hole, by definition, is an object whose escape velocity is equal to or greater than the speed of light. That means that anything that falls in cannot get out, because as far as we know nothing can exceed the speed of light.

Therefore, the first and most obvious danger from a black hole is, simply, falling in. If that happens, well, that's that. It's a one-way trip. You're done. End of discussion.

As a way a black hole can kill you, that's not terribly exciting—no death rays, no vast and terrible wreaking of havoc. Just bloop! And you're gone.

This lack of drama is a bit unsatisfying from a storytelling stance. But it also defies our common sense.* If you're in a rocket plunging into a black hole, can't you just turn the rocket around and thrust really, really hard and get out?

No, you can't. The extremely strong gravity near a black hole forces us to change the way we think about space, time, and motion.

Mathematically, the gravitational pull you feel from an object drops as the square of your distance from that object; double your distance from an object and the gravity you feel from it drops by a factor of $2 \times 2 = 4$. Get ten times farther away and the force drops by $10 \times 10 = 100$. Make the distance as big as you please; gravity goes on forever, and the force never actually drops to zero.†

So imagine you are on the surface of the Earth (which should be easy enough to do) and you have a ball in your hand. You throw it straight up into the air. As it goes up, gravity pulls on it, slowing its velocity. Eventually, the ball stops (velocity = 0) and then starts to fall back to Earth, accelerating the whole way down until you catch it.

Now imagine you throw the ball very high, like several miles high.

---

* Get used to that. Your common sense is going to take a beating here.

† This means that astronauts orbiting the Earth are *not* weightless because they are beyond Earth's gravity; they feel weightless because they're falling. When you are sitting in a chair, you feel gravity as pulling you down into the seat, which supports your weight. If there is nothing to support your weight, you don't feel the force of gravity, so when you're falling you feel weightless. This is why astronauts appear weightless in orbit (described as "free fall"). At the usual orbital height (300 miles or so) the Earth's gravity is only about 10 percent weaker than it is on the surface. Think of it this way: if the Earth's gravity weren't pulling on the astronauts, they would go flying off into deep space!

Gravity pulls it downward as it goes up, slowing it, but as it gets higher up, *the force of gravity is getting weaker because it's farther from the Earth.* So it's slowing down, but as it gets higher, the *rate* at which it's slowing is *itself* slowing, because gravity is getting weaker with height.

This means that if you can throw the ball at just the right speed, gravity will slow it down at the same rate that gravity itself is getting weaker. The ball will always slow down, but never actually reach zero. It will *always* move away from the Earth, but ever more slowly.

That's the definition of *escape velocity*—the initial velocity you have to give a projectile such that it will always move away from an object (like the Earth), always slowing down, but never stopping, and never falling back.

If you throw a ball up with slightly less than escape velocity, it will go a long way, but it will eventually come back. If you throw it harder, it'll just go away. At escape velocity—seven miles per second for the surface of the Earth—the ball is just able to escape from the Earth.

However, since gravity gets weaker with distance, the escape velocity gets smaller with distance too. If you were on top of a very tall mountain, the velocity at which you have to throw a ball is slightly less than the velocity you'd have to give it down at sea level. Also, escape velocity is an impulse; that is, it's the velocity you have to give an object all at once to get it to escape. If you can somehow continue to add velocity to a projectile as it heads up, then the concept of escape velocity gets a little trickier.

For example, you can in fact escape from the Earth by going more slowly than the escape velocity—at least, the escape velocity at the surface. Suppose you had a rocket with an inexhaustible fuel supply. You launch it at, say, 60 miles per hour, and keep the engines throttled so that it maintains that exact velocity, never slowing or accelerating. Eventually, it will be so far from Earth that the gravity is much weaker and the escape velocity has dropped to 60 mph.* At that point, you'll

* That distance is about 700 million miles from the Earth, so you'll be thrusting a long time: well over a thousand years. Better pack a lunch.

have escaped, despite never having gone anywhere near seven miles per second, the escape velocity from the *surface* of the Earth.

So, you might say, we can extrapolate this to black holes, right? If I fell into a black hole and had a big enough rocket, I could just thrust away, getting far enough away from the hole to where the escape velocity is something reasonable. Then I'm free!

Sadly, this won't work. If black holes were just another massive object then you'd be fine, just like the example above. But black holes are *not* just any old objects!

One of Albert Einstein's big breakthroughs in science was his idea that space is a *thing*. It's not empty; it's like a fabric in which massive objects sit. An object with mass has gravity, and that gravity bends space (the example in the last chapter was of a bowling ball sitting on the surface of a mattress, creating a dip in the middle). Any object going past a more massive one will have its path bent by that dip in space, by gravity.

IMPORTANT NOTE: Inevitably, when someone explains the idea behind black holes bending space, they use the analogy of a flat surface being bent by a heavy object, like the mattress and bowling ball. Unfortunately, this leads to a misconception that black holes are circles in space, surrounded by a funnel-shaped distortion of space. But that's not really the case: the reality is three-dimensional, and the analogy uses only two (the surface of the mattress can be considered two-dimensional but then is bent into the third dimension by the bowling ball). Black holes are spherical, and the bending of space is not shaped like a funnel. It's actually incredibly difficult to describe the shape of the space being bent, because we live in those dimensions, and describing them is like trying to describe the color red to someone blind from birth. We can describe it mathematically, make pre-*

---

* More specifically, *nonrotating* black holes are spherical. In reality, most black holes are created from rotating stars, and that rotation is amplified when the core of a star collapses into a black hole. Like any rapidly rotating object, black holes can bulge out at their equators from centripetal acceleration.

dictions about it, and possibly even use it to understand other aspects of physics, but picturing it in our heads is almost if not totally impossible.

So all the following descriptions of waterfalls, cliffs, and all that—those are analogies, two-dimensional representations of a warped three-dimensional reality. That may not make you feel any better, but the Universe has a way of making us uncomfortable. If that weren't true, this book would have no topic at all.

We now return you to the regularly scheduled death and destruction by black holes.

But a black hole doesn't just make a dip in space; it carves out a bottomless pit, an infinitely deep hole with vertical sides. Once you're inside, no velocity will ever get you out again. You fall in, and nothing can prevent it. For a black hole, the escape velocity at its "surface"—called the *event horizon*—is the speed of light.*

A more accurate way to think of this is using Einstein's mathematics and physics of relativity. Andrew Hamilton, an astrophysicist at the Department of Astrophysical and Planetary Sciences at the University of Colorado, Boulder, has studied black holes for quite some time, and has an interesting analogy:

> A good way to understand what happens is to think of a black hole as like a waterfall. Except that what is falling into the black hole is not water, but *space itself.* Outside the horizon, space is falling at less than the speed of light. At the horizon, space falls at the speed of light. And inside the horizon, space falls faster than light, carrying everything with it, including light. This picture of a black hole as a region of space-time where space falls faster than light is

* The event horizon is not the physical surface of a black hole. A black hole doesn't really have a surface; as far as we can tell, the matter in the black hole has shrunk all the way down to a mathematical point with literally zero size, called the *singularity.* The event horizon is the point where, at some distance from the singularity, the escape velocity equals the speed of light. The matter forming the black hole basically has no size, while the event horizon can be many miles across. Like I said, black holes are weird.

> not only a good conceptual picture . . . it has a sound mathematical basis [emphasis added].

This may seem like it breaks another of Einstein's laws—nothing can go faster than light—but that only applies to physical objects with mass (and light itself). Space itself is different than matter and light (another one of Einstein's Big Ideas) and so it can do whatever it wants, including moving faster than light.

If you are inside the event horizon, space is flowing down faster than light speed . . . and if you fall in, it's carrying you with it. If you try to paddle up a waterfall, you'll fail, because you cannot possibly get your boat moving *up* faster than the water coming *down.* So it is inside a black hole: with space flowing toward the center at transluminal speed, you can't paddle your rocket fast enough. You're doomed.

There is another way to think of this as well, but it's even weirder (if that's possible). If you look at the (fiendishly complex) equations that govern how space and time work near a black hole, you find that inside the event horizon, the variables representing space are constricted. Outside a black hole—like where you are now—you can move freely in space: up and down, front and back, left and right. However, inside a black hole, that freedom is removed. There is only one direction in which you can move: down.

Black holes are funny: even such a simple act as moving around turns out to be complicated. But the basic lesson is: if you fall in, no matter what, you're dead.

TIME OUT

Or are you?

Another one of Einstein's Big Ideas was that time and space are inextricably entwined, so much so that we actually refer to them together as space-time. When he formulated his theory of relativity, he realized that both space and time look different to someone who is moving relative to someone else. You may have heard of this already: imagine two people, each one in a separate spaceship, and each holding a clock. If one spaceship is moving very rapidly relative to the other, each of them will see the other's clock running at a slower pace, but their own will tick normally.

This is not a mechanical issue in the clocks; it's a physical manifestation woven into the fabric of space-time itself. And it's not just a guess: there have been countless experiments that show that Einstein was exactly right. Because space and time are two sides of the same coin, relative motion through *space* affects the way we perceive *time.*

Not only that, but gravity warps the way time flows as well. The closer you are to an object with strong gravity, the slower your clock will run—the slower time will appear to flow—as seen from someone farther away from the massive object. To you, your clock appears to be keeping time perfectly. Again, this has been confirmed via experiment. If you want to live longer, find the lowest spot you can! You'll experience more gravity, and others will perceive your biological clock as running more slowly. Of course, the effect is small for the Earth because its gravity is so weak. You might live a microsecond or two longer at sea level than if you lived your life out on a mountaintop, but that's about it. And worse, you yourself won't notice the difference, since you see your clock as running fine no matter where you are.*

But black holes have *lots* of gravity (and time to kill). Time dilation is very strong near a black hole. Imagine you are an astronaut near a black hole. You leave your copilot behind and let yourself drop in. As you approach, your friend, safe and snug in the capsule above, sees your time as flowing more slowly than his. The closer you get to the black hole's event horizon, the slower your time flows. You can try to talk to him, but your sentences get ssstttrrreeetttccchhheeeddd oooouuuuuttttttt . . .

---

* The corollary to this is: if you want to age less, move around really quickly all the time so others see your clock as running more slowly. Or you can sit around reading books on black holes and other astronomical dangers, and others will see your clock as running just like theirs.

When you fall into a black hole you are essentially riding along with space as it falls in. As you get closer, it falls in faster and faster. At the event horizon, space is falling into the hole at the speed of light. To a crewmate above, observing you through the light you emit, you never actually appear to cross the event horizon because the light you are emitting is going *upward* at the same speed space is traveling *downward.* It's basically treading water. As far as your crewmate can tell, you will remain suspended for an eternity at the event horizon, never falling in.

However, as a ticket to immortality, this is a bum ride. Because *this is only how your friend perceives it.* To your perception, you simply fall in. Plop! The event horizon, to you, is not a special place or time, and to you your clock takes that licking and keeps on ticking. You fall all the way to the center (to the singularity where all the matter is compressed to a dot), and you're dead.

Some people argue that because of this time-stretching, you can never fall into a black hole, but that's a misconception. You sure can, and when you do, you're gone. Your friends may not see it that way, but then they are sitting someplace safe while you're falling into a black hole, so who cares what they think?

PASTA-TA

In some ways, a black hole isn't all that different from any other object.

Anything that has mass has gravity. You do. I do. A bag of hammers does, the Earth does, the Sun does, and so does a black hole. The gravity you feel from an object depends on just two things. One is the mass of that object: double an object's mass, and the gravity you feel from it doubles as well.*

---

* Just to be clear, mass and weight are different. Mass is a property of matter; you can think of it as how much matter there is, and we measure it in grams or kilograms. Weight is the force of gravity on that mass, and we measure it in pounds. A cannonball has the same mass whether it's on the Earth or the Moon, but on the Moon it weighs one-sixth as much because gravity is one-sixth as strong; on the Earth 1 kilogram weighs 2.2 pounds, but on the Moon it weighs about 0.36 pound.

The other factor gravity depends on is your distance from the object—or actually, your distance from its *center of mass.* Remember, as described above, the force of gravity drops as the square of the distance, and that means the force *increases* at the same rate as you *approach* that object.

Let's take a look at the Sun. It's very massive—2×10^{27} tons (a 2 followed by 27 zeros), which is pretty impressive—and it's pretty big, about 860,000 miles across. If you could stand on the surface of the Sun without being vaporized, you'd feel a gravitational force about 28 times what you feel here on Earth.

But that's really the most gravity you could feel from the Sun. If you backed off (which is a good idea), the gravity you feel from it would drop, because you are farther away. And if you stand on its surface, you can't get any closer. If you did, you'd be *inside* the Sun. That would put you closer to its center, but now there is mass outside of your position, above your head. You can think of that mass as pulling you up, canceling a little bit of the gravity pulling you down.* As you get closer to the center of the Sun, the gravity you feel gets smaller. At the very center, you'd feel no gravity at all.

But now let's change the situation a bit. Let's compress the Sun so that the mass stays exactly the same, but it now has a diameter of, say, 3.6 miles. Since all that mass is now packed into a sphere only 1/240,000th as wide, the gravity at the surface will scream up . . . but the gravity you would feel 430,000 miles away (the original solar radius) *would be exactly the same!*

Think about it: the mass is the same, and your distance (from the center of mass of the compacted Sun) is the same. Since gravity only depends on these two things, the force from gravity that you feel is the same as it was when the Sun was normal-sized.

---

* Actually, because of the physics of solid bodies, the truth is that the mass above you doesn't pull on you at all, oddly enough. Newton was the first person to be able to work that out mathematically. Basically, once you are inside an object like the Sun, the only mass you need to worry about is from the stuff between you and the center.

The difference is, if you get closer, *the gravity goes up*. Before, it went *down* because you were inside the Sun. But now the Sun is small, so you can keep getting closer, and as you do, the force of gravity increases. It would go up and *up* and UP until you got 1.8 miles from the center (half the diameter), and at that point you'd be in real trouble.

Why? Because I didn't pull that "3.6 miles" number out of thin air. At that size, the Sun's gravity would be so strong that not even light could escape (you were wondering where this was going, I bet). That's right—if we could compress the Sun to that size, it would become a black hole.

The important point here is that from a long way off, the gravity from a black hole is exactly the same as from an object that's far larger but has the same mass. From a zillion miles away, the gravity from a black hole with ten times the Sun's mass would feel exactly the same as the gravity from a normal star with ten times the Sun's mass.*

Black holes are dangerous because you can get closer to them. That's where their real power lies. They are not necessarily more massive than other objects—many stars are far more massive than black holes. Their strength is in their *size*. Or their lack of it: they're *small*. They're so small that you can get really close, and their gravity increases enormously as you get closer.

This would have a very surprising consequence if you were brave—or foolhardy—enough to approach a black hole. A ripping good one, in fact.

If you fall into a black hole feet first, your head will be roughly six feet farther from the black hole than your feet (depending on your height, of course). Since gravity depends on distance from the center, the black hole will pull on your feet harder than it will on your head.

---

* No black hole formed in a supernova can have a mass less than about three times that of the Sun, though. The core of the exploding star has to be at least this massive or else it only forms a neutron star, not a black hole. So don't fret: the Sun cannot turn into a black hole.

From far away this difference in the force of gravity between your head and feet is small, but as you get closer it will increase.

This difference in force is called a *tidal force.** The Earth experiences a tidal force from the Moon: the side of the Earth nearer the Moon is pulled slightly harder by the Moon's gravity than the far side of the Earth. This raises a bulge on the Earth under the Moon. But counter-intuitively, it actually raises two bulges: the one *under* the Moon, and another one on the opposite side of the Earth, *away* from the Moon.

This happens because the Moon pulls harder on the center of the Earth than it does on the far side—the center of the Earth is closer to the Moon. So, in effect, the Moon is pulling the center of the Earth away from the far side; the result is a bulge on the far side of the Earth from the Moon. To an object suffering under tidal forces, it's as if it's being stretched—like taking one end of a rubber band in one hand and the other end in your other hand, and moving your hands apart.

Tidal force is similar to the force of gravity, but while gravity gets stronger with the inverse square of the distance, tides get stronger with the inverse *cube.* Halve your distance to an object and the gravity goes up by four times, but the tidal force goes up by eight times. Get ten times closer and the gravity goes up by a hundred times, but tides go up by a thousand times.

Obviously, this is going to be a problem.

Let's say you are an astronaut in a space suit, hovering over a typical black hole of, say, five times the mass of the Sun, which would have an event horizon about 18 miles across. Astronomers call this kind a *stellar mass black hole,* because its mass is about the same as a star's.† Let's also say you're a long way off, like 10,000 miles out. If you start your fall from this distance, the entire journey to the event horizon, even starting from

---

* Technically, this is a misnomer. It's not a force, but a *change* in a force. Unfortunately, this term stuck, and that's what we call it.

† Also, the diameter of a black hole is proportional to its mass; double the mass and the black hole's diameter doubles as well.

a standstill, will last only a couple of seconds! From that distance, the hole is pulling on you at an incredible 270,000 times the Earth's gravity. But oddly, you wouldn't feel it. Since you would be in free fall, with nothing to resist the force of gravity, you would actually feel weightless, just as skydivers do for the first few seconds of their fall, or astronauts as they orbit the Earth.

From this distance, the tidal force due to the six feet between your head and your feet isn't noticeable.

A second or so into your fall and you'd be accelerated even more. At 5,000 miles away, you have about one second left before you hit the event horizon, even from this distance. If you could speed up your reflexes, speed up your awareness (because you have only one second left to live, and we want you to be aware of what horrifying things are happening to you), you might notice an odd sensation, a feeling as if you're being pulled in two directions, toward *and* away from the hole simultaneously, as if people were playing tug-of-war, with you as the rope. The overall force on your body is still enormous, but the tides from the black hole generate a slight extra force on your feet of about a quarter of the Earth's gravity toward the black hole, and there will be an extra force on your head, up and away from the black hole, of the same amount. If you weigh 160 pounds, it would feel like a 40-pound weight hanging off your feet, and the same pulling your head up. It's uncomfortable, though not fatal. It will, literally though, make your hair stand up. Unfortunately, this changes a fraction of a second later.

At 1,500 miles away, the sensation is far stronger. It's as if you're being pulled apart like taffy—the force downward on your feet is now 10 Earth gravities, 1,600 pounds of weight. So is the force up on your head! The blood pools in your head, and you pass out (what fighter pilots call a "redout," the opposite of a "black out" when the blood leaves your brain). This, it turns out, is a blessing. You don't really want to be awake for the next few milliseconds.

That's when you get into real trouble. At 500 miles from the black hole, the opposite head-to-feet tidal forces are pulling you apart with a horrifying 550 Earth gravities, over 40 tons of weight. The human body isn't capable of withstanding that kind of stress. Soft tissue pulls apart, and your head and feet burst open from the pressure due to hundreds of pounds of blood pooling in them.

At 50 miles from the black hole surface, the tides are now over 700,000 times the Earth's gravity. It's like being suspended over an abyss with a cruise ship strapped to your feet. Your bones snap in half, and then again and again, pulled into tiny pieces.

But wait! There's more: you're not just getting stretched along your *length*, you're getting compressed across your *width*. Your left side falls toward the center of the black hole along a slightly different path from the one your right side wants to take. Both are trying to fall straight into the center of the hole, so your right side feels a force to the left, and your left side to the right. This squeezes you, and the force is also incredibly strong, about the same as the stretching force. You're being stretched out *and* squeezed in.

You're like a tube of toothpaste, and the black hole has a fist of steel. You're turning into a thin noodlelike tube of human goo.

When your feet—well, when what *used to be* your feet—are right above the black hole's event horizon, you're not even recognizable as a human being. You're stretched into an incredibly thin line, miles long, like a strand of pasta. Scientists call this process *spaghettification.**

And the black hole, as if in appreciation of the analogy, slurps you down.

So you see, simply *falling* into a black hole isn't the only way it can kill you. The journey there is half the fun.

FRIED BY THE LIGHT

As we saw in chapter 4, the birth of a black hole can wreak damage on an unimaginable scale, blasting out beams of radiation that will burn twin holes through the galaxy. The beams are generated when matter

---

* Yes, seriously, although a quick search didn't yield any mention of the term in professional physics and astronomy journals.

from the collapsing star forms an accretion disk around the black hole, channeling and funneling the matter into the hole. Coupled with the incredible magnetic fields involved, the beams are formed along the spin axis of the disk.

It turns out that this situation is not unique to the birth of a black hole. Anytime matter falls into a black hole, it can form such a disk, and beams can be generated as well. For example, if a black hole is orbiting a normal star (they form a binary pair, with one originally a very high-mass star that explodes and forms a black hole), then the hole can "siphon" matter off the other star. Usually this happens when

As matter falls into a black hole, it can pile up outside the event horizon, forming a flattened disk. Friction and other forces heat it to millions of degrees and can focus jets of energy and matter, as seen in this artist's illustration.

NASA/CXC/SAO

the normal star nears the end of its life and becomes a giant star (see chapter 8); the outer layers of the giant star can be drawn into the black hole.

The incoming matter forms an accretion disk just like when the black hole itself was born, and that disk gets incredibly hot. Surprisingly, the bulk of this heating is due to a rather mundane and everyday force: friction! When the matter gets near the black hole, it orbits faster and faster. Because of the ferocious gravity, a particle just slightly closer to the black hole can be moving substantially faster than one slightly farther out. They rub against each other, and friction heats them up.

As usual with black holes, they do nothing by halves. The friction can heat the disk to literally *millions* of degrees. Matter that hot generates radiation across the electromagnetic spectrum, from radio waves up to X-rays.* In fact, so much light is generated that, ironically, black holes (really, their accretion disks) can be among the brightest objects in the Universe.

Actually, this was how the first black hole was discovered. When the first X-ray satellites were launched, they found many such sources of high-energy light in the sky. One was traced to a giant star in the constellation of Cygnus. While this star, called HDE 226868, is a real bruiser—it has 30 times the Sun's mass—it just doesn't have the oomph needed to make X-rays on the scale observed. Sure enough, spectra† taken of the star indicated that it was being orbited by another object with about 7 times the Sun's mass, yet nothing was seen in the images. That meant it had to be a black hole; a 7-solar-mass star would have been be easy to detect. Plus, this naturally explained the X-rays blasting out of this system; the black hole (dubbed Cygnus X-1) was accreting matter from the giant star, and blasting out X-rays as the material swirled to its doom.

---

* Some black holes have been known to generate even higher-energy gamma rays as well, but this is due to nonthermal (not heat-related) processes.

† As a reminder, a spectrum is created when you break up light into its individual colors, which can tell you lots of interesting things about the object that emitted the light.

And by "blasting," I mean *blasting*. If you took all the energy emitted by the Sun and added it up, the black hole would be 10,000 times brighter *just in X-rays*. It's one of the brightest sources of X-rays in the sky, even at its distance of 6,500 light-years. If it were instead just a few light-years away, the X-rays could pose a threat to our satellites and manned space program (see chapter 2).

So you don't even have to be particularly close to a black hole for it to be dangerous.

Cygnus X-1 is the closest known black hole to the Earth, but by estimating how many stars are born capable of turning into black holes over the lifetime of the Milky Way, scientists have extrapolated that there may be millions more black holes in our galaxy alone.

I know what you're thinking: "Millions? *Millions* of black holes lurking throughout our galaxy? AIIIEEEE!"

Well, yeah. That sounds bad, so maybe we should take a moment and talk about that too . . .

JUST PASSING THROUGH

Our galaxy is lousy with black holes. They're everywhere! But what if one of them comes knocking on our door? Will that affect the planets, and even Earth?

Let's get this out of the way right now: this is an incredibly unlikely event. Space is big, and there's lots of room to knock around.

The Milky Way Galaxy is a collection of gas, dust, and something like 200 billion stars held together by their mutual gravity. It's a spiral galaxy, which means its major feature is a flat disk 100,000 light-years across punctuated by vast and beautiful spiral arms, like a pinwheel. To give you a sense of how big that is, the Sun, which is about halfway from the center to the edge of the disk, orbits the center of the galaxy at 160 miles per second, yet it still takes well over 200 million years to complete one orbit.

All the stars you see in the sky are relatively nearby; most are less than 100 light-years distant, a tiny fraction of the galaxy's size. The nearest known star is the triple system Alpha Centauri located a little over four light-years away. In English, that's about 26 trillion miles, so we're not exactly crowded out here in the galactic suburbs.

Over the Sun's lifetime there have certainly been stars closer to us than Alpha Cen, but that depends on what you mean by "close." Space is big, and stars are small. One study showed that a star passes about a parsec (3.26 light-years, or 20 trillion miles) from the Sun only once every 100,000 years, and that distance is still way too great for the star to affect us through its gravity. Closer encounters are even less common, and it would be unlikely in the extreme for a star to pass close enough for its gravity to significantly affect the Earth.

And that's for stars in general. There are thousands of stars for every black hole in the galaxy. I hope you're getting a sense that a close encounter with a black hole has pretty low odds. The closest one known is the aforementioned Cygnus X-1, which is a pretty distant 1,600 light-years away. That's a bit of a bummer for a book with a title like this one's, but we must face reality, even if it means we're safe from the black hole menace.

Still, as with the other topics in these chapters, it's fun to think about. What would happen if a black hole came to pay us a call?

There's a good chance we'd never see it. If it was traveling solo through space, it would just be, well, a black hole in space, invisible, emitting no light. A stellar mass black hole is typically only a few miles across, making it far too small to spot until far too late.*

Although nearby stars are orbiting the center of the Milky Way in

---

* Because black holes curve space, light gets bent as it travels near one—think of it as a road going around a curve, and a car on that road having to follow the curve too. An approaching black hole might be detected through this distortion—we might see star positions apparently changing, and bigger background objects like nebulae and galaxies getting smeared out. But this distortion is small when the black hole is far away, and likely to escape our notice until it is inside our solar system. And while this might give us decades of warning, there's not a whole lot we could do about it short of evacuating the Earth . . . which presents its own set of issues.

pretty much the same direction and with roughly the same speed as the Sun, there is some variation. Like cars around a racetrack, a small difference in speed means some cars pass each other. Even though the cars may be traveling at 200 mph, they pass each other relatively slowly, at a few miles per hour. The same is true for stars. The Sun is orbiting the galaxy at 160 miles per second, but so are other nearby stars. The typical speed at which we see them move relative to the Sun is far less, just a few dozen miles per second. At those speeds, it would take years for a star to get from the orbit of Pluto to the orbit of the Earth.

But, it turns out, there are exceptions. Some stars are real speed demons, and interestingly, we see some compact objects like neutron stars moving across the galaxy at amazingly high speeds, hundreds of miles per second faster than you'd expect.

These runaway stars were pretty mysterious at first, but it's now thought that their high velocities are the product of the supernova explosion in which the stars themselves were born. If the supernova event itself is slightly off-center, exploding more to one side than the other, the material and energy blown out of the star will act like a rocket, pushing it in the other direction. Incredibly, even a slightly off-kilter explosion can impart vast energies to the neutron star remnant, accelerating it to high speeds. Also, if the supernova progenitor is in a close binary system, orbiting another star, the orbital speeds can be several hundred miles per second. When the progenitor explodes, both stars get flung away in opposite directions at large velocities.

Either way, it's physically possible, even likely, that a neutron star or black hole can be slicing across the galaxy at a pretty good clip.

What if it's aimed at us? Will we survive a drive-by of a 10-solar-mass black hole, moving at, say, 500 miles per second (a large but reasonable velocity)?*

---

* At lower velocities, which are in general much more likely, the same events would unfold, just more slowly.

A black hole getting near the Earth would be bad enough, but if it was actively "feeding," gulping down material, the outpouring of X- and gamma rays would cook our planet to a crisp.

DANA BERRY, SKYWORKS DIGITAL INC.

The scenario at the start of this chapter should give you a taste of what's to come. But it depends, of course, on how close it gets to the Earth. Let's run through what happens on approach and see.

As the marauding black hole approaches the solar system, a planet will feel its gravity as well as the gravity from the Sun. As the hole gets nearer, the planet feels its gravity getting stronger. Like a toy being pulled on by two greedy kids, the planet's orbit will start to distort. If the passage is distant enough (say, it's on the opposite side of the Sun), the planet may be relatively unaffected—its orbit may become a bit more elliptical, but that's about it. But if the hole gets close enough, its gravity will dominate over the Sun's, especially for more distant planets like Uranus or Neptune, where the Sun's gravity is relatively weak. If that happens, the planet may start to orbit the black hole, or, more likely, the hole's gravity will simply slingshot it out of the solar system. In general, in an encounter where you have two massive bodies (like a star and a black hole), and a smaller one (a planet) gets involved, the smaller one is very likely to be ejected from the system.

Such is the scale of disaster of which we are talking: whole planets are literally flicked away.

As the black hole approaches the Earth, we surface dwellers won't really notice any change in gravity at first, but the Earth as a whole will. Its orbit around the Sun is perturbed more and more as the black hole nears. When the hole is about three times farther away from the Earth as the Sun, or roughly 300 million miles, its gravity will equal that of the Sun.* When that happens, the Earth is no longer "bound" to the Sun. It could fall into the Sun, or fall into the black hole, or be ejected from the solar system.

Which do you prefer? Hmmm . . . no happy endings here.

Not that we'd have much of a choice. And things are about to get a lot worse.

The tidal force from the black hole, responsible for the spaghettification of our unfortunate astronaut earlier, will begin to affect the Earth as well. At a distance of 300 million miles, where its gravity is equal to that of the Sun, the tidal force is about a third of the Sun's. That's not much (much less than the tides from the Moon), and unlikely to cause any damage.

But the black hole is nearing by 500 miles every second, 40 million miles every day. At that speed, it can cover those 300 million miles in about a week, so just a day or so later its tides start to dominate. By the time it's the same distance as the Sun from the Earth, its tides will be

---

* The force of gravity drops as the square of distance and goes up with mass. When the hole is $\sqrt{10}$ or about three times farther away from the Earth as the Sun, its gravity is $\frac{10}{(\sqrt{10})^2} = \frac{10}{10} = 1$ times the Sun's gravity.

five times stronger than the Moon's. Water will flood coastal communities, and small earthquakes may be felt.

A day later, it's half as far as the Sun. Its tides are now 40 times that of the Moon. Tidal waves* many yards high inundate the coastlines, killing millions of people. And every minute the force gets stronger.

Just a few hours later, when the black hole is a mere 7 million miles away (30 times farther away than the Moon), someone standing on the surface of the Earth will feel the same force from the black hole as from the Earth itself. For just a few moments, you'd be weightless, and a small jump would send you flying upward.

Enjoy it while it lasts. At that distance, the tides from the black hole are a staggering 20,000 times that of the Moon (well, what *used* to be from the Moon—it would have already been ejected from orbiting the Earth by the black hole's mighty gravity). The Earth is under colossal strain, and earthquakes would be larger than any ever measured. Whole continents would begin to tear apart, and volcanic eruptions would be constant.

Finally, the tides are more than the Earth itself can handle. It gets torn apart, spaghettified on a planetary scale. What's left of our once lush planet is shredded and heated to millions of degrees, finally spiraling into the maw of the black hole.

And that, once again, is pretty much that.

Amazingly, all this time, the black hole itself is so small—just under 40 miles across—that even if it weren't totally black, it would still appear as nothing more than a dot in the sky. Only the most powerful telescopes would see it as anything else . . . but again, it's black. There's nothing to see.

As for the prognosis for the rest of the solar system, it depends on the trajectory of the black hole. The Sun itself may escape relatively unharmed if the hole doesn't get too close to it—otherwise it'll get torn up pretty well. If the black hole misses by a sufficient margin, the Sun's

---

* Earthquake-induced floods are called *tsunamis,* which many people erroneously call tidal waves. In this case, we are literally talking about a wave caused by tides.

path around the galaxy might be only slightly affected, and the Sun itself may survive.

Isn't that comforting?

BLACK HOLETTES

The smallest black hole that can form in a supernova is about twelve miles across, and that's pretty scary. Picture it this way: it's about twice the size of Mount Everest, and three *quadrillion* times the mass.

That's terrifying! But if big is scary, is small cute?

When it comes to black holes, no. They're all pretty frightening. But can smaller black holes even exist?

Theoretically, they might. Called *primordial black holes* (or mini black holes, or sometimes even quantum black holes), these would be very small, with masses much less than those of their stellar mass cousins, and maybe even less than the Earth's. They've never been observed, but there may be countless examples of them floating in the depths of space, and they're called primordial because they'd be as old as the cosmos itself.

In the very early Universe, just moments after the Big Bang, vast energies and densities were being tossed around like snowflakes in a blizzard. Space itself was folded like origami, and for the briefest of instants, just a razor's edge of time after the initial Bang, conditions were such that a relatively small amount of matter could find itself squeezed by immense forces. If the density of the matter shot high enough quickly enough, it would actually form an event horizon and become a black hole. These mini black holes could have had very modest masses, on the scale of the mass of mountains, a few billion or trillion tons.

Such a tiny black hole would be weird, even for a black hole. The event horizon would be teeny-tiny: a black hole with the mass of the Earth would be only about half an inch across—the size of a marble. One with the mass of an asteroid or a mountain would be far smaller than an atom!

Obviously, such a black hole would be even harder to detect than the normal flavor, which may be why they've never been seen (although, to be honest, they may not exist at all; they're still theoretical). Even if they were to accrete matter, the flow onto a mini black hole would be so small that they'd be invisible even from relatively small distances.

But mini black holes have a secret. You might think that black holes always grow, eternally eating matter and energy, getting larger in the process. But black holes, it turns out, may not be forever. They may evaporate.

In the 1970s, the scientist Stephen Hawking had an idea. It was pretty crazy, but when you're dealing with black holes, ideas reach the "crazy" category pretty quickly. By applying the laws of quantum mechanics and thermodynamics to black holes, he realized that in some sense, black holes have a temperature. They can actually radiate away energy, just as normal matter does. That energy has to come from someplace, and as he conjectured, it comes from the black hole mass itself.

Here's how it works. In quantum mechanics, the rules by which the Universe plays get truly bizarre. Energy and mass are interchangeable, with energy easily able to be converted to mass and vice versa.* But another odd aspect is that space itself can belch out small amounts of energy out of nowhere, ex nihilo if you will. In fact, the fabric of space is positively bubbling with energy that can pop out into the real world.

This may seem to violate one of the most basic properties of the Universe: you cannot create or destroy energy or matter. Normally that's true. But this energy created out of nothing can exist for only very brief amounts of time, *as long as it goes away,* back into the nothingness whence it came, very quickly.

It's like borrowing money from the bank. Eventually, you have to return it. And the more you borrow, the faster you'd better pay it back.

If the Universe decides to belch out a tiny bit of energy, that's okay, as long as it quickly goes back into the fabric of space. All laws of nature are conserved if this happens quickly enough.

* We ran across this before—it's how stars make energy from fusion, via $E = mc^2$.

But if it happens near the event horizon of a black hole, things get sticky. The gravity of the black hole can cause this bundle of energy to fragment, creating matter. This happens in the bigger Universe all the time; gamma rays, a form of energy (light), can convert into matter if they collide with each other or interact with matter. Because of the way things must balance, two particles are created: one is normal matter, like a regular old electron, say, and the other is antimatter. Antimatter is exactly like matter, but it has an opposite charge, so an antielectron (called a *positron*) has a positive charge. That counteracts the negative charge of the electron, and the cosmic ledger books remain balanced.

But if this happens right at the very edge of the event horizon, it's possible that one particle can fall in while the other remains free. It can escape, and to a distant observer it looks as if the black hole has emitted a particle. This mass (or, equivalently, energy) balance must be repaid, and it comes out of the mass of the black hole. In effect, the black hole has lost a tiny amount of mass.*

Another way to look at it is using tidal force. The particles appear—poof—near the black hole event horizon. The tidal force from the black hole pulls the two particles apart. One falls in, and the other gets out. It takes energy to separate the particles, which has to come from somewhere. It comes from the black hole itself—energy and mass are equivalent, remember, so the black hole loses a tiny bit of mass when this happens.

This process is very slow, and depends on the mass of the black hole. The lower the black hole's mass, the smaller the event horizon, and the easier it is for this process to happen (or, equivalently, the lower the mass the stronger the tides are near the event horizon). Since the black

---

* You might think that the particle that fell in balances the mass from the particle that escapes, so the black hole has lost no net mass. However, because of the laws of gravity (and how weird they get inside a black hole), inside an event horizon a particle can actually have *negative* energy—essentially, the black hole holds on to it so tightly that the total energy of the particle is less than zero. This balances the positive energy of the particle outside the black hole, and everything remains even, except for the energy lost in separating the particles. Remember what I said earlier about common sense?

hole is radiating away mass and energy, this whole process acts as if the black hole has a temperature—it's warm, and it emits energy to cool off. The smaller the black hole, the higher the temperature, since it loses mass and energy more rapidly. This means, in turn, that massive black holes will last longer than smaller ones, since they radiate away their mass more slowly. A stellar mass black hole will have a temperature of only about 60 billionths of a degree!

But a smaller black hole will be "hotter," radiating away particles more rapidly. As it loses mass, its temperature goes up, and that means it radiates away matter even faster . . . it's a runaway process, accelerating all the time. Once it gets below a certain mass—about a thousand tons—it releases all the remaining energy in less than a second. Kaboom! You get an explosion. A *big* explosion: energy and matter would scream out of the black hole, releasing the equivalent of the detonation of *a million one-megaton nuclear bombs.*

A mini black hole created in the formation of the Universe with a mass of about a billion tons would be just about at that stage now. Any with smaller masses would have evaporated long ago, and more massive ones are still stable. A stellar mass black hole can tool along for incredible lengths of time before worrying about evaporation; the projected life span of such a hole is more than 10^{60} years, which is far, far longer than the current age of the Universe (but see chapter 9 to find out what happens when that time finally arrives).

No quantum black hole explosion has ever been seen (though for a while, some people conjectured it might explain gamma-ray bursts), but even that amount of energy would be difficult to detect from light-years away. Could quantum black holes wander the galaxy? What would happen if one got too close; would it be as dangerous as a stellar mass black hole?

Imagine a black hole with a mass of 10 billion tons—roughly the same as a small mountain—heading toward Earth. It is far too small to detect through its distortion of background stars—it's less than a trillionth of an inch across, smaller than an atom. The gravity from it wouldn't be enough to affect the planets, the Moon, or the Earth, which are far, far more massive. However, we'd certainly notice it long in advance: because of Hawking radiation, it would burn fiercely at a temperature of billions of degrees! Because it's so small, it would actually be fainter than the faintest star you can see with your unaided eye, but satellites like NASA's Swift observatory might detect the gamma rays it emits as it approaches.

Finally, it dives through our atmosphere. It wouldn't draw in much matter as it fell through the air; a 10-billion-ton black hole would hardly be noticeable gravitationally even from a few yards away. But up close, at distances less than an inch, the gravity would be hundreds of times that of the Earth. Any air within that distance would get sucked right in. This might form a small and temporary accretion disk, but at typical collision speeds of several miles per second there would hardly be time for it to do much before plunging into and beneath the Earth's surface.

To such a black hole, the solid matter of the Earth might as well be a high-grade vacuum. Far smaller than an atom, it would pass right through the Earth, and at supersonic speeds it wouldn't get much of a chance to eat much matter. It would almost certainly be traveling faster than Earth's escape velocity too, so it would blow right through us and move on, perhaps just the teeniest bit heavier, and then continue on its merry way.

Well, that's not very dangerous. And not much fun either. Let's try a bigger one.

Suppose instead we have a black hole with a mass equal to the Earth itself, and, through an unfortunate series of circumstances, it was headed right for us. Moreover, just to make sure we get some fun results, let's also assume it's moving very slowly relative to the Earth, only a few miles per second. This is incredibly unlikely—it probably wouldn't happen once even if the Universe were a thousand times older—so it is really just a "what if" scenario, and you needn't let it keep you up at night.

Getting such a slow approach would be hard, but not impossible. For example, if it was moving slowly enough to start with, and it swung by a planet or two and the Moon on its way in, the primordial black hole's orbit could be changed sufficiently that it would be able to collide with the Earth and not keep traveling out into space. This would be quite the gravitational dance, and less likely than, say, sinking every pool ball on the opening break ten racks in a row. But we're looking for some action here, so let's see what this gets us.

Things would be . . . interesting. First off, we'd never directly detect its approach. Hawking radiation from it would be very weak; its temperature would be similar to that of space itself, far below zero, so it would not be emitting any observable light. However, we'd certainly see it indirectly. As it approached, we would experience vast tidal forces. The black hole is very small—about half an inch across, the size of a marble—but has the mass of the entire Earth. From far away, remember, the force of gravity is the same as the Earth's. The Moon would be affected profoundly; most likely it would be ejected from the Earth's orbit forcibly. It's possible, if things were just right, that the Moon's velocity relative to the Earth would be slowed enough that it would plummet toward us like the giant stone that it is. If it impacted, the least of our worries would be the black hole. The energy released in the impact would vaporize the surface of the Earth and kill every living thing on it down to the base of the crust.

While that's quite the apocalyptic scene, we want the black hole to do the deed in this fantasy scenario, so let's assume that the Moon gets ejected. What happens as the black hole approaches the Earth?

When it is still 240,000 miles away, the same distance from the Earth as the Moon, its tides would be huge, 80 times the strength of the Moon's. As it gets closer the tidal force strengthens, prompting earthquakes and floods.

Eventually, it falls into our atmosphere. At that point, while it is, say, 100 miles above the Earth's surface, the destruction would be beyond comprehension. Just the gravity alone would be awesome: you'd feel a force upward, toward the black hole, 1,600 times stronger than Earth's gravity! Anyone within sight of the black hole's approach would be picked up and flung away like a leaf in a tornado.

As it plunged through our atmosphere it would suck down quite a bit of gas, possibly creating an accretion disk and emitting high-energy radiation. There would be an enormous shock wave, similar to a nuclear detonation, which would wreak all sorts of havoc—if there were anything left to be wreaked upon.

When the black hole reaches one mile above the ground, anyone still standing (not that there could be) would feel a tidal force of 40,000 times Earth's gravity trying to rip him apart. Spaghettification would be inevitable. Everything on the Earth's surface would be literally torn apart.

When the black hole hits solid ground a moment later, the accretion rate would increase, heating it up considerably. There might even be enough energy emitted quickly enough to act like an explosion . . . but at this point that's fairly moot.

To the black hole, which is incredibly dense, the Earth is essentially a vacuum. It would fall pretty much freely through the Earth. Its ferocious tides would tear the planet's surface apart as it fell, most likely destroying everything above.

In a sense, that's too bad. We'd miss the *really* scary part.

The black hole is so dense that it would essentially be orbiting the center of the Earth *inside* the Earth itself. As it passed through the Earth's matter, even a microscopic chunk of rock would feel a tremendous change in the force of gravity if it got too close to the black hole, easily equaling millions of gravities. This tidal effect would tear the rock to bits, heating it up hugely, vaporizing it. Because of this, the black hole deep inside the Earth would be surrounded by a sphere of intensely hot and incredibly compressed gas, similar to what you might find in the core of the Sun. At the center of this cloud, the black hole would be greedily swallowing down the matter. As the black hole moved through the Earth it would be like a blowtorch, heating the material around it and feeding on it.

Even though the black hole is small, this vaporous halo is big enough that it would rub against the solid or liquid rock around it, creating friction. This friction drags on the black hole, which over time slows its speed through the Earth. It would spiral in, falling to the Earth's core. There, the pressure of the overlying matter would give it a continuous source of food . . . and *it would eventually eat the Earth.*

The whole planet.

Nothing would be left . . . except the black hole.

We'd be long gone by the time that happened, of course. But to an observer off planet, those last few moments—only a few decades after the black hole first approached the Earth—would be spectacular. The shrunken and distorted planet would be only a few meters across, and white-hot. Finally, in a millisecond's time, the last piece would fall into the black hole's accretion disk. Heated to millions of degrees, the remaining bits of what was once our planet would probably explode outward as they absorbed the tremendous energy emitted near the black hole's event horizon. When the debris cleared, there would be nothing left to see, just a slightly larger black hole, now a whole inch across after its gorging, calmly orbiting the Sun.

MAN HOLE

While those last scenarios are certainly apocalyptic—they're the first ones we've run across where the Earth is quite literally destroyed—they're also by far the least likely to occur. We don't even know if primordial black holes exist, for example, or in what numbers if they do. And even if they are out there, and in huge numbers, the odds of one getting close enough to the Earth are incredibly low. And even though we are very sure that stellar mass black holes lurk in the galaxy, the odds of one of those getting close enough to ruin our day are microscopic as well. Space is vast, and the Earth is tiny, so we're pretty easy to miss. The very fact that the Earth has existed for about 4.6 billion years is rock-solid proof of that.

But what if one doesn't start in the depths of space? What if one were to start off right here, *on* Earth?

The new generation of particle colliders—what used to be called atom smashers—can actually slam subatomic particles into each other so hard that it's theoretically possible that they will create extremely tiny mini black holes. A few years back, this news made some headlines when it was revealed that the Relativistic Heavy Ion Collider (RHIC) in New York might be able to do just that. Would the Earth get eaten by an artificial black hole?

Many newspaper articles speculated it might, but there are two reasons why that can't and won't happen. One is that, as we saw, tiny black holes will evaporate through Hawking radiation extremely rapidly. A black hole made by the types of collisions done at RHIC would last the tiniest fraction of a second. They'd never get a chance to accrete any mass before evaporating (and their mass would be so small that the explosion would be really tiny too).

Second, the energies created at RHIC are actually much smaller than what naturally occur at the top of the Earth's atmosphere billions of times a day! Cosmic rays—subatomic particles accelerated to fiendish energies in supernova explosions—slam into the air all the time at far higher energies than we can hope to create here on Earth. These are more than enough to create extremely tiny black holes, yet here we are. Over the billions of years that these particles have been raining down on us, not once has the Earth been eaten by a subsequently created black hole.

Newspapers, magazines, and TV like to inflate such stories because they know they will sell. But when you look at the actual science, you see that we're in no danger of being gobbled up by a black hole, whether by nature's hand or our own.

And that is the hole truth.*

---

* Actually, it's *not* the whole truth. Black holes still may have something to say about our eventual fate; see chapters 8 and 9.

CHAPTER 6

Alien Attack!

*M**INDLESSLY—AT LEAST, LACKING WHAT WE WOULD** call a mind as we know it—it examined the bright light of the star ahead of it. Employing a highly sophisticated complex of observational instrumentation, it patiently took data, examining each bit of information as it came in. After weeks of steadily staring at its target, the results were in.*

The star was orbited by several gas giants. Each of these had icy moons with possible water under the surface. The star also had not just one but three smaller planets with the potential for liquid water as well. And on the second one of these out from the star were unmistakable signs of biotic life—free O_2 in the atmosphere at large levels of disequilibrium. If it had been equipped with emotions it would have whooped with joy. Instead, it silently and efficiently began preparing for the next phase of its mission.

Using sophisticated engineering and technology the probe began to slow its approach to the solar system. Its fantastic speed—nearly that of light itself—gradually bled away over the course of nearly a year. Course corrections were made, angling the probe this way and that. All the while, it took observations, scouting for the target it needed. Finally, the target was acquired: a metallic asteroid over a mile wide. As the probe passed the asteroid, aiming carefully, it released a small package just a few meters across.

The package was a probe in its own right, and it used its onboard rocket to decelerate further and land on the surface of the asteroid. It immediately sent an "all clear" signal to the mothership, which did not respond, but instead sharply accelerated away, heading off to the next star on its list, a star it would not reach for decades.

On the asteroid's surface, a hatch opened on the probe, and a small spider emerged . . . then another, and another. In all, a dozen such robots started crawling over the landscape. Composed of a sophisticated amalgam of metal, ceramics, and spun carbon fiber, they went to work: digging, smelting, manufacturing. They worked without fatigue, without emotion, tirelessly day and night (such as there was on the slowly rotating asteroid). After a month they were ready.

Like a fungus expelling spores, the asteroid erupted in thousands of tiny explosions. Each puff imparted velocity to a ball of metal a meter across, each of which headed toward one of the planets and moons initially targeted by the interstellar probe. Inside each ball were over a hundred of the spiders. The robotic arachnoids were possessed of sophisticated programming, but in the end the goal was simple: convert any and all available materials into more spiders. When enough were manufactured, build more mothership probes. Launch them, and repeat the cycle.

They needed metal of almost any sort, and what they couldn't find they could create. Their programming was very sophisticated, honed over millions of years of such missions. And they weren't picky about materials; almost anything would do. Each spider could create the components needed to replicate itself in just a few hours, and these would then move on to replicate themselves as well. Once the first spiders touched down, they could cover a planet in just a few days, converting everything—everything—*into more spiders and more probes.*

Being smaller and closer, the surface of Mars was destroyed first. The rocks were rich in iron, which made things easier. Within hours, that many more spiders went off in search of raw material.

Food.

Earth fell within days. The first spiders landed in Australia and consumed everything in their sight. Rock, metal, gas; all could be converted if needed. Water, plants, flesh—these would do as well. Humans never had a chance. Though the intense light of the interstellar probe's engine had been tracked by Earthbound telescopes for months, it didn't answer any hails, and there wasn't enough time for any of the governments to react anyway. By the time the spiders landed it was already far too late. They swept over the planet, and after less than two weeks there were in essence no living creatures left on Earth. The entire surface of the globe had been converted into robotic factories. Within a year, bright flashes blossomed over the planet as more interstellar probes were launched, each an exact replica of that first one—which itself was generations removed from the first probe launched so many eons before. That first probe was long since dead, having expended all its packages. It was no longer useful. But its progeny "lived" on, sweeping across the galaxy.

And now, several thousand more headed out into deep space. When the original mission started, mankind didn't exist; only hominids ambled across the African plains. Their descendants ruled the Earth, but their reign was brief. All those billions were now gone, converted into hordes of little metal spiders and more interstellar probes.

Man's dream of reaching the stars was finally achieved, but not quite in the manner in which he thought.

THAT'S LIFE

No matter where you go on the Earth, you find life.

On the plains, on mountaintops, high in the air, down to the deepest ocean depths, life abounds. Even far underground, microscopic life has adapted to conditions we would consider lethal. Life is everywhere.

Earth seems marvelously tuned to support life, but that's an illusion: *we* are the ones who are in fact tuned by evolution, as are all the other forms of life on, below, and above the Earth's surface. As the Earth has changed over the eons, so has life. It seems almost inevitable that, once life first got its start on Earth, it would flourish.

We know there are other planets in our solar system, and even orbiting other stars. If life is so plentiful here, it stands to reason it may also be on those other worlds. They may teem with simple microorganisms, and it's possible that there could also be more complex forms of life in space, things that we would recognize as intelligent.

If that were so, what would they think of us? Would they be a threat to us?

To understand that, we'll start by taking a short journey backward in time—though my definition of "short" may differ a bit from yours.

A BRIEF HISTORY OF THE SOLAR SYSTEM

Some 4.6 billion years ago, you were spread out over countless of cubic miles of space.

So was I. So was the book you're holding now, and the clothes you're wearing, and everyone and everything you've ever known, ever seen, ever touched, ever dreamt about. All your atoms were part of a vast disk, tens of billions of miles across and a million miles thick. The disk was almost entirely made up of hydrogen and helium, but it also contained scattered impurities of zinc, iron, calcium, phosphorus, and dozens of other elements. It rotated slowly, held together by its own gravity and the gravity of the lumpy swell at the center.

Over millions of years matter accumulated in the center of the disk, gravity pulling in ever more material. As it compressed it got hotter, and eventually the core temperature reached 27 million degrees Fahrenheit. Hydrogen fusion triggered, and it was at that moment that our Sun became a real star. Light flooded out, followed by a wave of subatomic particles, a nascent solar wind.

In the meantime, the outer parts of the disk were busy accumulating matter as well. At first clumps only stuck together because of chemical processes. Ice crystals formed farther out from the Sun, where temperatures were low. Chunks of silicates smacked into each other and stuck. Over time, as the aggregations grew, so did their mass, and so did their

Self-portrait, 4.6 billion years BC. This illustration shows what our solar system looked like when it was young. A disk of rock, gas, ices, and metals revolved around the newly born Sun, just starting to form the planets we know today.

NASA/JPL-CALTECH/T. PYLE (SSC)

gravity. These *planetesimals* started actively pulling in more matter in a runaway process—more mass, more gravity, more matter, more mass, and so on—that was finally only quenched when there was no more material to accrete. Some of these new planets were small, some large. Some decent-sized ones were ejected out of the system entirely when they passed too close to the larger ones.

The ones that survived all had rocky cores and thick atmospheres. Some had atmospheres thousands of miles deep, and no real surface to speak of. Others were smaller, with dense atmospheres to be sure. They also had molten surfaces, heat left over from the formation process.

When the Sun in the center turned on, its fierce light and solar wind hit the new planets. The pressure from the light and the ejected matter slammed into the thinned disk, blowing away the leftover detritus, clearing the space. Eventually, all that was left was a handful of planets, billions of asteroids, and trillions of icy comets, all orbiting a young, hot star.*

Our solar system was born.

It looked different then! Jupiter was farther out from the Sun than it is now, while Saturn, Uranus, and Neptune were closer in. Their mutual dance of gravity would eventually migrate them to their current distances. In the inner solar system, Mercury, Venus, Earth, and Mars all had thick, soupy atmospheres. Over time, as the inner planets solidified, they too would change. Mercury, so close to the Sun and with such low gravity, would have its atmosphere stripped away. Also, the tiny planet's lack of a magnetic field left it open to the full brunt of the Sun's solar wind, which aided in tearing Mercury's atmosphere away atom by atom.

Venus would eventually lose its hydrogen and helium, but chemical processes over billions of years, including a runaway greenhouse effect, would give it a thick atmosphere of carbon dioxide. Trapping the heat from the Sun, the planet would become a forbidding desert of kilnlike heat. The surface rocks are always just a hairbreadth from being molten.

Earth too had a dense atmosphere, nothing at all like today's—it looked more like Jupiter or Saturn back then, consisting mostly of hydrogen and helium left over from the disk from which it formed. At its distance from the Sun, incoming heat (plus the heat coming from the surface) puffed up the thick air, in a delicate balance with gravity. Over

---

* A relatively recent idea is that giant planets like Jupiter and Saturn may have formed from the direct collapse (called *fragmentation*) of material in the disk, rather than being built up by collision. This scenario is gaining some ground among astronomers, but the actual birth mechanism of planets is still somewhat debatable.

millions of years, the lighter elements were lost, leaving behind an atmosphere of carbon dioxide, water vapor, carbon monoxide, ammonia, methane, and other noxious gases, most of which leaked out from inside the Earth.

Eventually the surface cooled, forming a thick crust over the molten, semiplastic mantle of rock. The heavy elements like iron, iridium, and uranium sank to the center. The radioactive elements decayed, generating heat, adding to the heat trapped inside left over from the formation of the planet. Convection currents began, a magnetic dynamo ensued, and Earth became protected from the ravages of the solar wind.

Not that the young Earth was safe from threats from space. A lot of the material in the solar disk was swept up by the planets, but not *all* of it. A huge repository of asteroids still roamed the system, and their paths would sometimes intersect those of the planets. Shortly after the planets formed, they were all bombarded mercilessly. Nearly every solid surface in the solar system bears witness to this devastation; a quick glance at the heavily battered and cratered surface of the Moon will confirm it.

The Earth bore the brunt of its share of collisions too. It was hit significantly more than the Moon was, being bigger and having stronger gravity—in fact, the most commonly accepted theory on the formation of the Moon itself is that it coalesced from material ejected when the Earth was hit by a very large object, perhaps as big as Mars, an apocalyptic collision that is terrifying to imagine. But over billions of years plate tectonics and erosion have wiped out all the evidence of this early bombardment. Only the most recent craters are still around; even ones older than a few million years are nearly invisible. Still, this early pelting from space created an immensely hostile environment. Anytime things began to settle down, some fifty-mile-wide rock would come crashing down, resetting the geological clock.

Eventually, though, the rain of iron and rock ceased. As the Earth cooled, more complex molecules could form. The methane in Earth's atmosphere was a source of hydrogen, as was ammonia, which also had nitrogen. Carbon dioxide provided the carbon, which, when liberated from the oxygen, could combine to form ever more complex chains of atoms. Amino acids—the building blocks of proteins—probably formed quite early, and started combining in new and interesting ways. Lightning from the atmosphere and ultraviolet light from the Sun may have provided energy needed to break up and reform the molecules. At some point—no one knows exactly when or how, but it was virtually simultaneous with the cessation of the bombardment from asteroids—the molecules formed into a pattern that had a fantastic property: it could reproduce itself. As things stand today, this molecule was probably incredibly simple, but it still possessed the amazing ability to gather up raw materials and construct them in such a way as to reproduce a copy. These then went forth and multiplied.

It was hardly more than a simple chemical reaction, or really a long series of them. These reactions needed materials—the elements found in the air and surface—and emitted waste products. One of these waste products was oxygen. As oxygen built up in the air, the chemical processes started to change. Oxygen, to many of these very simple microbes, was toxic, as waste products sometimes are to the organisms that produce them. As the gas built up, it poisoned them. Some species of microbes adapted to the new environment (and their descendants still live today as blue-green algae and other forms of life), but those that couldn't perished; they had thrived on Earth for millions of years, but their own waste killed them.* However, a slightly different complex molecule was around at the same time. It was able to use the waste. Oxygen, when combined with other chemicals, can release a lot of energy, which in turn can be useful for reproduction and increased metabolism. This different microbe fed off the waste of the others, and when the oxygen levels got high enough that the first life-forms started dying, the oxygen-users were ready for the coup. They took over. Some oxygen-producers survived, mutating and adapting all the time as

---

* If you wish to view this as a cautionary tale, be my guest.

some fit into the environment better than others. Oxygen-users that were more adept at using the fuel flourished; others died off.*

Asteroid impacts, vast solar flares, and the random nearby supernova may have wiped out this process or culled it back to near-extinction levels many times over the millions of years the scenario played out, but eventually a toehold (or a pseudopod hold) too firm to shake off was established.

Earth became alive.

Now, this tale is but one way life may have arisen on Earth. We don't know for sure how it happened. We're not even sure where it happened: land, sea, air, in the deep ocean . . . or even on Earth. Our planet is only one of several where, in the early life of the solar system, conditions were ripe for the development of life.

MARTIAN CHRONICLES

Mars, though, was smaller than the Earth, and farther from the Sun. It cooled more rapidly than the Earth did, and may have gone through the same series of events, but on a shorter time scale. We don't know this for a fact, of course, but it's certainly possible that Mars had a thriving microbial ecosystem long before the Earth did. Unfortunately, it was doomed. Once Mars cooled down enough, it lost any magnetic field it might have had, so it became victim to the Sun's solar wind. Its weaker gravity allowed almost its entire atmosphere to leak away to space, and now the pressure on the surface is a miserable 1 percent of Earth's, thinner than at the top of Mount Everest.

Robotic explorers sent to Mars have shown us a clear history of a

---

* Before all this oxygen was exhaled by the new microbes, the Earth's atmosphere contained a large amount of methane. Oxygen combines readily with methane, so most of the atmospheric methane was destroyed when oxygen became abundant. Methane is a strong greenhouse gas, so when it disappeared the Earth may have cooled significantly, gripping the planet in a global ice age. This would have aided any mass die-offs of bacteria as well.

The Mars rover Opportunity took this snapshot of the Red Planet on March 1, 2004. The presence of sulfates and other chemicals in the rocks indicates that water once flowed over the surface of Mars.

NASA/JPL

watery surface, however. Chemicals in the ground and rippled wave patterns indicate that in the past there were vast floods, perhaps as frozen water underground was heated by volcanism or impacts. There are also indications of ancient lakes, now desiccated, as big as the Great Lakes in the United States. Even today there are (still controversial) hints of transient, short-lived events where liquid water flows on the surface . . . only to quickly evaporate into the thin air.

But four billion years ago it was a different story. Did ancient Martian oceans teem with simple life-forms, bacteria, protozoans? We don't know, and we may never know. Still, future probes may yet find fossils in ancient Martian rocks.

But suppose for a moment that Mars once was alive. What does that have to do with Earth?

In 1984, an unusual meteorite was found in the Allan Hills region of Antarctica. Meteorites are relatively easy to find there; on the ice, a black rock sticks out pretty well. The dry air preserves the meteorites well too.

This particular meteorite, dubbed ALH84001, packed a huge surprise. The chemical composition of the rock itself was very similar to the known composition of the rocks on the surface of Mars. Inside it were small bubbles of gas, trapped within the rock when it formed. When the ratios of the various elements of the gases were measured, they matched the ratios found in the Martian atmosphere! Scientists checked this against the chemicals in other planets as well, but no other gas ratio in the solar system matched nearly as well as that of Mars.

Clearly, ALH84001 was an interplanetary interloper, a rock from the Red Planet. It wasn't even the first discovered; many were found before their now more famous cousin. How these rocks got here is a matter of some irony.

The surface of Mars bears the marks of a violent history. Like every other planet in the solar system, Mars has been bombarded over its life by asteroids and comets. In contrast to Earth, with its thick atmosphere and watery surface, erosion works more slowly on Mars, so we still see the craters, scars from the ancient impacts.

When such a collision occurs, rock from the surface is launched into the air, of course. But some of that material can be given so much energy that it can actually leave the planet altogether and fly into space. From various studies of ALH84001, it has been found that it formed very early in the history of the solar system, cooling from volcanic lava some 4.5 billion years ago. It sat on or below the Martian surface for almost all the time since, and then suffered a terrible blow. An asteroid impact on Mars hurled the rock into space. It orbited the Sun for at least 16 million years, judging from the rate of cosmic-ray impacts on its surface. Its orbit got nudged this way and that every time it passed by its home planet, and eventually the orbit changed enough that it began to move closer to the Sun. Some 13,000 years ago, our planet got in the way. The rock fell to Earth, got lodged in the Antarctic ice, and sat patiently waiting to be found.

So the rocks from Mars that landed on Earth as meteorites were themselves launched into space by impacts from much larger rocks.* This is a very violent way to get a rock into space, of course, and you'd expect that an impact like that would destroy any rocks, or certainly damage them considerably.

But recent studies have shown that that might not be the case. A low-angle impact, for example, can loft rocks relatively softly (though it's not exactly the kind of ride for which you'd line up for tickets). There may be other factors involved as well, including pressure waves from the atmosphere and under the rock due to the impact that also combine to soften the blow.

However it worked, very small structures inside ALH84001 survived the ordeal. When the rock was examined by scientists, they found a wealth of intact structure inside, including some that indicated that the rock had been exposed to flowing water at some point in its past. When they took microscopic images, however, they got the shock of their lives.

Tiny wormlike formations were evident inside the Martian rock. They looked for all the world like fossilized bacteria! In 1996, a press conference was held, and the scientists who examined the rock announced what they had found, indicating that it could be, for the very first time in the history of the world, evidence of life outside the Earth. They were very careful to say that the evidence was not (pardon the expression) rock-solid, but very compelling. The "fossils" were actually the weakest of their evidence, and they took great pains to say that they might not be fossils at all. They might be natural formations caused by any number of processes.

---

* It's much more difficult to get meteorites from the inner planets out to Earth because they would have to fight the gravity of the Sun as well as that of their own planet. Despite that, some meteorites tentatively identified as being from Mercury have been found.

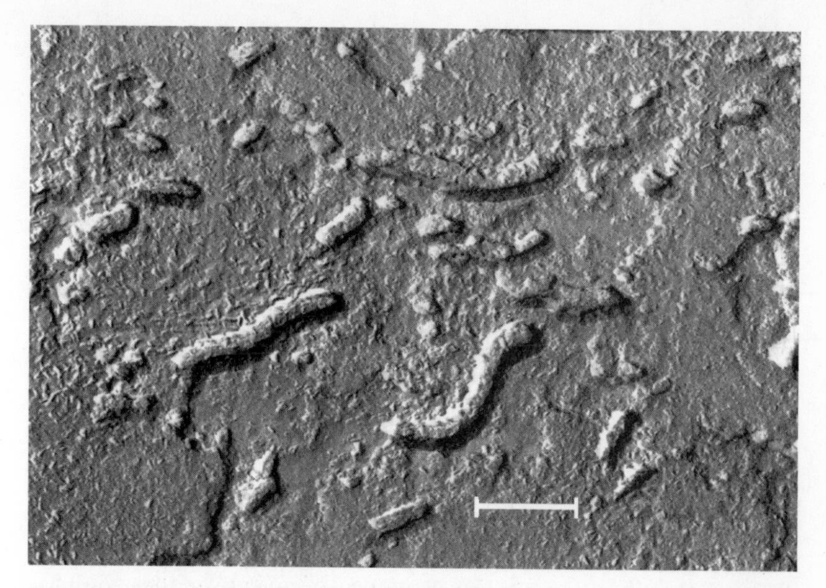

A microphotograph of the "Mars rock" ALH84001 shows the presence of small, wormlike objects. While they look like primitive life-forms similar to those on Earth, their origin is unclear. They are far smaller than any similar terrestrial organism. The scale bar is 0.5 micron across; for comparison, a human hair is 50 microns in width.

DAVID MCKAY, JOHNSON SPACE CENTER

Of course, the press ran with the fossil interpretation; a picture, after all, is worth a thousand words, and will sell a million issues of a newspaper. It was quite the media sensation. Over the years, however, one by one, the evidence for life in the rock has come under fire. At the moment, the best one can say is that the evidence is interesting, perhaps even still compelling in some ways, but everyone agrees we will need far better sources before we can talk clearly about ancient life on Mars.

THERE ARE THOSE WHO BELIEVE LIFE **HERE** ***BEGAN*** **OUT THERE . . .**

But this brings up an intriguing possibility: if life did originate on Mars first, it could have been brought here by the same mechanism that brought us ALH84001. Is it possible that life on Earth actually came from Mars?

At first blush, this sounds like a dumb idea. Earth is flourishing with life—it's actually quite hard to find any place on the planet that isn't infected with it—but Mars is quite dead. Still, the steps needed to seed life on Earth are at the very least possible: life may have originated there first; a plausible mechanism exists for it to have gotten here; and conditions here eventually were good enough for that life to take hold.

The idea that life was brought to Earth from space is called *panspermia.* It's a fascinating topic, fraught with one simple problem: how do you prove it?

Honestly, I'm not sure you can.* But it's very hard to rule it out. How do you perform experiments to test it? Re-creating conditions that haven't existed in billions of years can be tough, and even then it doesn't prove things one way or another because of uncertainties inherent in the experiments. But experiments along those lines can steer thinking in directions that may lead to further progress; in science, a good experiment is worth a thousand suppositions.

An interesting test would be to look for fossilized microbes on Mars that have a chemical tie to early life on Earth. A clear example of RNA-based or DNA-based fossil bacteria would be incredibly compelling

---

* Panspermia is studied by many solid researchers with good reputations, but like any other field of science at the cutting edge of knowledge, it suffers from its share of kooks. Like UFO believers, there are people who point at everything they find as evidence of panspermia, from red-tinted rain in India to odd microbes found floating in the upper atmosphere. After looking into these cases, I have found them to exhibit the same problems as every other pseudoscientific claim: lack of solid observations, poorly controlled experiments, shoddy research, a lack of critical thinking, and a very strong tendency to jump (and leap, and catapult) to conclusions. We may yet find strong—even solid—evidence of life from space, but it will be uncovered using scientific methods: careful observations, reasoned experiments, and judicious thinking. Otherwise you just get cold fusion: a lot of pomp, but no circumstance.

evidence in favor of panspermia—either life started on Mars and came here, or both Mars and Earth were seeded from some third source.*

Until such evidence is found, we can only conjecture.

Still, in principle, it's possible to examine the processes involved with panspermia.

The step after getting the life-laden rock off Mars (or some other body) is the journey here. ALH84001 spent at least 16 million years in space, and possibly more, where it was exposed to the hard vacuum of space, bombarded by high-energy subatomic particles, and bathed in killer ultraviolet rays from the Sun.

Anything that could survive that would have to be pretty tough.

Microbes *can* be tough hombres. Some bacteria can form protective spores around themselves, shielding them from the ravages of heat, cold, drought, and radiation. One type of bacterium—*Deinococcus radiodurans*—can survive intense doses of radiation, hundreds of times what's needed to kill a human. It is rather like a computer with multiple file backups: it has many copies of its DNA that it can use in case some get destroyed by radiation, and the tools it uses to repair its own DNA (every cell nucleus has a repair kit) appear to be extremely adept at dealing with extreme conditions.

Of course, it would also help if any microscopic stowaway on board a meteoroid were gift-wrapped carefully too. A rock dislodged by an asteroid impact and flung into space would be irradiated by all manner of destructive sources. But if the rock were large enough, it might protect any microscopic cargo. Cosmic rays may not penetrate very far into the surface, for example. Other disruptive influences like ultraviolet light from the Sun, particle irradiation from the solar wind, and the odd solar flare or coronal mass ejection would have a hard time getting deep into the rock as well. Some early experiments in lofting bacteria samples into space and exposing them to the environmental

---

* As noted before, getting rocks from Earth to Mars is possible, but considerably more difficult and therefore much less likely.

hazards there indicate that some microbes could survive for a period of time in space.

If some Martian protovirus or bacterium were to wend its way deep into a rock that got blasted off Mars, then there is some chance—small, but finite—that it could survive the journey.

It would also have to survive the journey through our atmosphere. But again, if the rock were large enough only the outer layers would burn off as it plummeted through the Earth's air. If the meteoroid disintegrated into smaller pieces above the ground, the individual impacts wouldn't be so jarring to its biological stowaway either. A small rock would simply plop down, and if it fell into water or mud, which seeped into cracks, the microbe might suddenly find itself getting, in a literal sense, food delivery.

It's important to note that Mars isn't the only source of potential life. Comets are giant balls of rocks and ice that orbit the Sun, and are known to contain rather complex organic compounds, some of which are precursors (or have at least basic chemicals needed) for life. It's possible that comets impacting the young Earth brought much of the water to our planet, and brought these chemicals as well. A meteorite that fell in Australia in the 1960s was also found to have amino acids in it, including glycine and alanine, both of which are commonly found in animal proteins. Even giant clouds of gas and dust in space are found to be rich sources of complex organic compounds. One study done by scientists even showed that DNA or RNA from bacteria, if shielded well, could actually survive being blown by the solar wind to *another star.** That work is pretty speculative, but it shows in principle that transport across large distances, even vast ones, is theoretically possible, even if very unlikely.

Incidentally, it should be noted that a comet need not directly impact the Earth to transfer any contents. When a comet gets near the Sun, the frozen material sublimates (turns directly to a gas) and leaves

---

* This is easier if the wind is actually from a red giant; those kinds of stars emit far less UV radiation that can damage or kill the bacteria.

the comet, forming the long tail. If the Earth sweeps through a comet's tail, the cometary materials can mix with the Earth's atmosphere. It's still a somewhat violent process, since the velocities are so high, but in principle the comet stuff can reach the Earth relatively intact.

Let's also be careful here and state explicitly that all of these are the building blocks of life, and not life itself. But the fact remains that the components needed for life as we know it not only exist in space but are relatively abundant—and these sources made it down to the ground intact enough to be studied by scientists. It's entirely possible that space is simply buzzing with life, and it's also possible that this is where terrestrial life got its start. If true, actually *proving* it would be one of the most colossal and fundamentally profound discoveries ever made.

But it also could mean trouble. If life exists out there as microbes, and some of them came to Earth now, could there be a less than happy ending? It's all well and good for the surviving bacterium, and if panspermia is true, we owe our existence to space bugs. But that was more than three billion years ago.

What would happen if this event were to repeat itself today? We've all seen the movies *The Blob* and *The Andromeda Strain*. Could an interplanetary infection invade us, wiping out humanity (or mutating us into horrible nasty gooey things)?

To be honest, probably not. Life here is pretty tough, and anything coming down from space will have an uphill battle trying to take us over. It's my opinion that they won't be able to win. But the fight itself depends on what type of gooey thing is on the prowl.

VIRAL ADVERTISING

Viruses, for example, are a favorite in science-fiction scenarios. There may be millions of types of viruses on Earth; we have only scratched the surface in investigating their diversity. Most are actually extremely simple structures: they are a snippet of DNA code wrapped in a protein shell called the *capsid*. They cannot reproduce on their own; instead, they invade a cell, inject their DNA into the cell's nucleus, and then urge it to copy the viral DNA. Viruses are the stealthy ninjas of the submicroscopic world, sneakily invading the factory of the cell and turning it against itself.*

When enough viruses are created, they burst out of the cell, rupturing and destroying it. They then scurry off, seeking out more cells to subvert. If the body cannot fight off the infection, and the virus is virulent enough, the host can be killed. The tissue of the body literally liquefies.

Nasty.

There are other types of viruses too. Some use RNA, not DNA. Others attack bacteria and not tissue cells. And not to make you uncomfortable or anything, but your body is currently *brimming* with these viruses. Most are completely harmless. Some do cause a variety of issues—for example, they can throw off your body's ability to regulate its systems, resulting in illnesses from mild to severe—but most don't kill. They have to attack in ferocious numbers to do that, or be particularly virulent, like Marburg (which has a mortality rate of about 25 percent) and the more famous Ebola (with its truly terrifying 80 to 90 percent mortality rate).

The structural simplicity of viruses is both a blessing and a curse when it comes to an invasion from space.

Because they are such simple structures, viruses are resistant to many of the problems a more complicated microbe might have with exposure to space. Prolonged periods of vacuum, low temperatures, and even some radiation may not prove an obstacle to them. Embedded deep in a rock, they could fall to Earth intact, only to be opened like a cursed pharaoh's tomb by a hapless scientist.

But if they got under his skin, they might starve to death.

That's because viruses are generally adapted to attacking *one specific kind* of organism. A virus that can infect a plant can't harm a butterfly, and one that is adapted to attacking bacteria (called *bacteriophages*)

---

* In fact, viruses are so simple that many scientists don't consider them to be alive. Their lack of ability to reproduce on their own substantiates that (plus they don't eat or excrete in any real sense either).

can't hurt a human. Viruses are too simple to change radically, and the DNA or RNA snippet in the virus is like a key to a lock. A car key won't work in a house door.

So even though any hypothetical space-borne virus might survive all the way into the lab of a scientist, it's incredibly unlikely that it would swarm and multiply and turn us all into raging zombies.*

So in reality, viruses aren't a big threat. They would find us completely incompatible for their purposes, and would quickly die out.†

Score one for life on Earth.

BUG IN THE SYSTEM

Interplanetary bacteria are another horror movie staple, and while they have some advantages as invaders over viruses, they're also unlikely to be much trouble to us Earthbound creatures.

Unlike viruses, which are programmed to fit certain types of cells or proteins, bacteria are less choosy. And while viruses use our own cells' machinery against us, bacteria consider us more as a flophouse. Like an unwanted guest, they can eat your food, mess up the place, and, of course, overstay their welcome.

The main difference between a virus and a bacterium is complexity. Bacteria are cells in their own right, and are considered alive. They can

---

* It's possible that viruses were the precursors of life on Earth. They certainly have been around a long time, and coevolved with us. Even if life on Earth got its start from space viruses landing here via panspermia and kick-starting our ecosphere, such viruses would be harmless today. We've evolved for a long time since then, and it's not terribly likely they will still find a lock to fit their key.

† Incidentally, there is an even simpler structure than viruses, called *prions*. They aren't much more than complex aggregations of proteins, and aren't actually alive in any real sense. They can, however, mess up the structures of normal proteins in tissue, causing large holes to form in cells. This in turn produces all manner of horrifying problems, such as convulsions, dementia, and death—mad cow disease and scrapie in sheep are caused by prions. However, like viruses, they can attack only certain types of proteins, and any prions that evolved on another world are unlikely in the extreme to be able to infect terrestrial life.

ingest food, excrete waste, and reproduce on their own. Give them a warm, wet environment with the nutrients they need, and they'll do all three of these functions with abandon.

Our bodies make excellent sites for what bacteria need. Our bodies are loaded with bacteria, as with viruses. Bacteria are in your gut, in your skin, and living on your eyelashes. They're everywhere, and in fact it's been estimated that there are ten bacteria in your body for every single human cell!

You're outnumbered.

The vast majority of bacteria inside you are benign. They either don't do anything harmful or exist in numbers too small to do any damage. Many are beneficial to us; without them we'd die. They help us digest our food, for example; they also create vitamins and boost our immunity to more harmful types of bacteria. They even help us digest milk.

An excellent example of such a bacterium is *Escherichia coli,* more commonly known as *E. coli.* This little ovoid bug lives in huge numbers inside your intestines, and has the underappreciated job of helping you process your waste matter.* Normally, they live happily in your gut, doing whatever it is they do. But that's not always the case. *E. coli* eats and poops too, and some strains of the bacterium exude a toxic chemical brew. In low doses your body can handle it. But if you get too much in your system, it can make you quite ill. Food poisoning, for example, can be caused by eating food that has been contaminated by *E. coli.* If the infection is bad enough, it can be fatal. *E. coli* can also get out of your intestines (through a hole or herniated region) and into your abdomen, causing peritonitis.

The list of possible problems bacteria can invoke is lengthy (diarrhea, vomiting, nerve damage, cramps, fever . . . you get the picture), but usually we live in an uneasy truce with the bugs inside us.

The bacteria inside us have, of course, evolved along with us so they can maintain this symbiotic relationship. A hypothetical bacterium

---

* I'm trying to be polite here. Cut me some slack.

that evolved on Mars, say, or some other planet would not enjoy this luxury. Still, the effect of an alien bacterium on us really depends on what it needs, and, pardon the expression, what it excretes.

If all it needs is a warm place with water and some nutrients, then any port in a storm, as they say. Your intestines will look just as good as any other place. And if the bacterium multiplies, and the colony emits a toxin, then that can be trouble.

But is that likely?

In reality, almost certainly not. The very complexity that makes bacteria more versatile and therefore more adaptable than viruses is also their Achilles' heel: it makes them more fragile. Their internal machinery is unlikely to survive the journey through space, and their entry into our atmosphere.

Moreover, the conditions alien bacteria need to survive would make Earth look pretty unfriendly. The chemistry of the surface of Mars, for example, is very different from Earth's. Its thin atmosphere means it's hit by a high level of UV light from the Sun. There is very little if any water on most of the surface, and some readings indicate rather high levels of hydrogen peroxide, a chemical that tends to destroy terrestrial bacteria (which is why it's used to clean wounds, though it should be noted that it is produced by some forms of life on Earth—notably the bombardier beetle, which uses it to ward off predators). Any bacterium that evolved to survive on Mars would most likely find Earth to be a very difficult environment—too wet, too hot, too *alien*.

Of course, not all life on Earth likes the same things we do. Some bacteria like extreme cold, some like it hot, others eat sulfur, and some like the extreme pressures found deep underwater or underground. These *extremophiles* are abundant on Earth, and may exist on Mars as well. But even if they are there, deep under the Martian surface, it's unlikely they'd get scooped up and carried away by an asteroid impact.

Looking to other worlds in the solar system for potential bacterial breeding grounds is even more futile. Europa, a moon of Jupiter that is covered in ice, may harbor a vast water ocean under its surface. It's an excellent candidate to look for life beyond Earth, but it's a low-probability location for anything that'll think our environment is cozy. The ice on Europa is probably ten miles or more thick; any impact that could loft a subsurface ocean-dwelling microbe into space would also be powerful enough to vaporize said microbe.

Another potential home for life is Titan, one of Saturn's moons. Titan is aptly named: it's over 3,000 miles in diameter (about the size of Mercury) and sports a thick atmosphere of nitrogen, argon, and methane. It rains there, but the drops are liquid methane! It's *cold* on Titan, about −300 degrees Fahrenheit. Any water on the surface is frozen into a solid harder than terrestrial rocks. And while biochemists have speculated that life could arise in such a weird environment, it would be utterly alien to us. Any bug capable of living there would find itself in the equivalent of a blast furnace on Earth.

It seems that as incubators go, we've struck out of potential bugs in the solar system. Any alien microbes that would have evolved for Earth-like conditions almost certainly wouldn't survive the trip.

Of course, this assumes that any form of life Out There is just sitting back and waiting for a ride. Maybe, though, the more sophisticated types would prefer to drive.

WHERE ARE THEY?

The question was asked so succinctly by the physicist Enrico Fermi in the early 1950s, over lunch with some other scientists. They were discussing the recent spate of flying saucer sightings and considering interstellar travel, human or otherwise. When the topic turned to aliens, Fermi asked, "Where are they?"*

The question, simple though it is, has a rich backstory. The basic idea is that by now either we should have detected intelligent life in our

* The exact quotation is lost to antiquity; it may have been "Where is everybody?" which is just as pithy.

galaxy or it should have come visiting. Since neither has occurred,* asking where the aliens are is a reasonable thing to do.

Let's assume that for aliens to come knocking, their circumstances must be something like ours: Sunlike star, Earthlike planet, development and evolution of life over billions of years, discovery of technology, then the capability to travel between the stars. How likely is all this to happen?

For that we can turn to the Drake Equation. Named for the astronomer Frank Drake, it categorizes all the necessities of advanced life and assigns probabilities to them. If you fill in all the terms correctly what pops out is the number of advanced civilizations in the galaxy (where "advanced" is defined as being able to send signals into space—which is how we'd know they're out there).

For example, the Milky Way Galaxy has roughly 200 billion stars in it. About 10 percent of these stars are like the Sun: similar mass, size, and so on. That gives us 20 billion stars to work with. We're just now learning how planets form around other stars—the first planet around a Sunlike star was discovered in 1995—but we're finding that stars like the Sun are rather likely to have planets. Even if we assign a ridiculously low probability of there being planets around another star (say, 1 percent), there are still hundreds of millions of stars out there with planets. If we assign a ridiculously low probability of these planets being Earthlike (again, say, 1 percent), then there are still millions of Earthlike planets. You can continue to play this game, estimating how many planets can support life, how many have life, how many have life capable of technology . . . each step in the chain is a little less firm than the last, but even the most pessimistic view of this series indicates we shouldn't be alone in the galaxy. The estimates of the number of aliens out there vary widely, literally from zero to millions.

---

* I discount UFO sightings. Despite a zillion blurry photos, obvious fakes, and shaky video, there has not been a single unequivocal piece of evidence that we have been visited by aliens, *ever.* Deal with it.

ARE WE ALONE?

That's not terribly satisfying, of course. The lower estimate is sobering. Maybe, just maybe, we really *are* alone. In all the galaxy, in all the vast trillions of cubic light-years of emptiness, ours is the very first planet to harbor creatures that can ponder their own existence.* This is a humbling and in some ways frightening possibility. And it's possibly true.

Another possibility is that life might be common, but "advanced" life is rare. Books have been written on this topic, and it makes for an interesting argument. Maybe once life gets to a certain stage, it tends to go navel-gazing and never develop or care about technology (alien psychology is a difficult topic to get too deeply into). And I hope that by the time you get to this point in this book, I've made it clear that civilization-ending events occur uncomfortably often over geologic time scales. Maybe every civilization eventually gets wiped out by some natural event before it can develop space travel advanced enough to prevent it.

I don't think that's a good answer, actually. We are within years of being able to prevent devastating asteroid impacts on Earth. We know we can properly shield ourselves from solar events. Our astronomy is good enough to pick out nearby stars that might explode, so if we saw one ticking away we could devote ourselves to getting away from it. All of these advances are quite recent, happening in a blink of the eye compared to how long life has existed on Earth. It's almost impossible for me to imagine a civilization intelligent enough to explore the heavens yet not advanced enough to preserve its own existence.

TALK IS CHEAP

I'm suspicious of the other end of the estimates of the Drake Equation as well, that there are millions of aliens out there as advanced as we are or more. If that were true, I think we'd have unequivocal evidence of them by now.

---

* There *is* another way to be alone, as we'll see in a moment.

Remember, besides being vast, the galaxy is *old.* The Milky Way is at least 12 billion years old, and the Sun only 4.6 billion. If we imagine a star like the Sun forming just 100 million years earlier—a drop in the bucket compared to the age of the galaxy—then it's not hard to imagine an alien civilization rising many millions of years before humans did. We know that life arose easily enough on Earth; it got started as soon as the bombardment period ended and the surface of the Earth calmed down enough for long-term growth of life to occur. This implies strongly that life takes hold given the smallest opportunity, which in turn means it should be abundant in our galaxy. And, despite a list of disasters epic and sweeping, life on Earth has managed to get this far. We are intelligent, we are technologically advanced, and we are a space-faring species. Where will we be in a hundred million years?

Given that stretch of time and space, an alien species really should have knocked on our door by now.

They should have at least placed a call. Communicating across the vastness of space is easier than actually going there. We've been sending signals into space since the 1930s. These are relatively faint, and an alien would have a hard time hearing them from more than a few light-years away, but we've leaked out stronger signals as time has gone on. If we wanted to target a specific star, it's not hard to focus an easily detectable radio signal to any star in the galaxy.

The reverse is true as well: any alien race with a strong urge to chat with us could do so without too much effort. The Search for Extraterrestrial Intelligence (SETI) is banking on just that. This group of engineers and astronomers is combing the sky, scanning for radio-wave signals. They are almost literally listening for aliens. The technology is getting so good that the astronomer Seth Shostak estimates that within the next two dozen years, we'll be able to examine the million or two interesting star systems within a thousand light-years of Earth. This will go a long way toward our discovering whether we are alone or not.

The one drawback with SETI is that the conversations will tend to be a bit boring. If we detect a signal from a star that is really close on the galactic scale, say, a thousand light-years away, the dialogue will really be a monologue. We'd receive the signal, reply, and have to wait two thousand years for them to get back to us (the time it takes for our signal to reach them plus theirs to come to us again). While SETI is a great and worthwhile endeavor—and if they find a signal it will be one of the most important events in the history of science—we're still used to thinking of aliens actually *coming here.* Face-to-face, as it were, assuming they have faces.

But a thousand light-years is a long way (6,000,000,000,000,000 miles). That's quite a hike, yet it's practically in our laps compared to the size of the Milky Way.

Is that why we haven't been visited? Maybe the distances are simply too great!

Actually, not so much. A trip to the stars wouldn't take that long at all, if you maintain a sense of scale.

TO BOLDLY GO

Let's assume that we humans suddenly decide to fund the space program. And fund it *really* well: we want to send probes to other stars. That's no easy feat! The nearest star system, Alpha Centauri (which has a Sunlike star and is worth a look-see), is 26 *trillion* miles away. The fastest space probe ever built would take thousands of years to get there, so we couldn't really expect a payoff in the form of pretty pictures anytime soon.

However, that's the fastest probe ever built *so far.* There are ideas out there on the drawing board that would make much faster unmanned probes, even ones that can move at a goodly fraction of the speed of light. Some of these include fusion power, ion drives (which start off slowly but accelerate continuously over years, building up ferocious speeds), and even a ship that explodes nuclear bombs behind it to provide a huge impulse in speed.* These methods can drop the trip time from millennia to mere decades.

* This is serious: called Project Orion, it was studied in the 1960s. The acceleration isn't smooth—getting kicked in the seat of your pants by a nuclear weapon gener-

This might be worth doing. It's expensive, sure. But there are no *technological* barriers to this idea, just social ones (funding, politics, etc.). Let me be clear: if we had the will, *we could build spaceships like these right now.* In less than a century we could be sending dozens of interstellar emissaries to other stars, investigating our own neighborhood in the galaxy.

Of course, the trip times and the actual construction of the fleet make it difficult to explore much real estate. The galaxy has billions upon billions of stars, and building that many starships is impossible. Sending one probe per star isn't cost-effective. Even if we let the probe simply sweep through a star system on a fly-by, moving on to the next star, exploring the galaxy would take forever. Space is big.

But there's a solution: self-replicating probes.

Picture this: an unmanned spaceship from Earth arrives at the star Tau Ceti after a journey of fifty years. It finds a series of small planets and begins its scientific observations. This includes a census of sorts—taking measure of all the bodies in the system, including planets, comets, moons, and asteroids. After some months of surveying, the probe will move on to the next star on its docket, but before it leaves, it sends a package down to a particularly promising nickel-iron asteroid. This package is in fact a self-starting factory. Once it lands, it mines the asteroid, smelts the metal, refines out the necessary substances, and then autonomously *builds more probes.* Let's say it builds just one probe that, after a few years of construction and testing, blasts off for another star system. Now we have two probes. A few decades later they arrive at their destinations, find appropriate accommodations, and then go forth and multiply again. Now we have four probes, and the process repeats.

The number of robot ambassadors builds very rapidly; it's exponential. If this takes exactly one hundred years per probe, by the end of a millennium we have $2^{10} = 1{,}024$ probes. In two millennia there are a million probes. In three thousand years there will be more than a

---

ally isn't—but it can build up tremendous speed. Unfortunately, the Nuclear Test Ban Treaty (chapter 4) forbids the testing of such a spaceship.

billion. Now, in reality, it's not that simple, of course, but even a pessimistic approach shows that we can explore every single star in this galaxy in something like 50 million years, maybe a bit less.

Well, that sounds like a long wait! And we're still a long way off from being able to do it. The technology is formidable.

But hang on—remember that civilization we considered, just 100 million years in advance of us? Given that much time, they could easily have examined every single star in the Milky Way, looking for life. If they saw our warm, blue world, one would think they'd make some note of it. Still, it's possible they came here 50 million years ago and missed us humans (mining the Moon for a monolith à la *2001: A Space Odyssey* maybe isn't as silly as it sounds), or maybe they just haven't gotten here yet.

But given the time scales, that seems unlikely. It just doesn't take that long to map out a whole galaxy and visit the appropriate planets. That's why I don't think the "millions of civilizations" number from the Drake Equation is correct. We'd have seen them by now, or at least heard from them.*

IN SPACE NO ONE CAN HEAR YOU SCREAM

But sometimes I wonder. Given all this information: the likelihood of life, the relative ease of galactic exploration, the time spans involved . . . and the fact that we have not detected any other life in our galaxy at all, there is another possibility that is worth mulling over.

---

* This logic means that a *Star Trek*–like galaxy—where there are lots of aliens at roughly the same technological level—is extremely unlikely. If life abounds in the Milky Way, civilizations are far more likely to be separated by gulfs of millions of years. Some of the aliens will be more like Q and the Organians (hugely advanced beings in the *Star Trek* universe), with one or two like us, and the rest not much more than extremely primitive microbes and yeasts. Another *Star Trek* aspect of this is the Prime Directive: the procedure to quarantine rising civilizations until they develop the capability of interstellar travel. That's an interesting idea, but I don't buy it: it means that *every single* alien species out there will obey it. It only takes one maverick to spoil the secret.

Consider: what spurs technological advancement more than anything else?

War.

The first Cro-Magnon who beat an opponent over the head with a tree branch was also the one who was most likely to live a bit longer, and be able to reproduce. The army with rifles will (in general) beat the one with spears. The country with missiles will (in general) beat the one with cannons. The ones with electronic remote-controlled drones, spy satellites, and instantaneous communication will outmaneuver the ones without.

Nothing advances technology like good old-fashioned aggression. Even one of the most noble events in human history—man walking on the surface of the Moon—was initiated because of the cold war, the space race with an enemy vast and powerful. Americans imagined Soviet missile bases in orbit and on the Moon, and the motivation to beat them was in place.

When I was a kid, it was fashionable in more intelligent science-fiction stories to assume that any aliens we meet were bound to be friendly—no warlike race would be able to get their act together long enough to reach the stars.

Humans are on the verge of falsifying that statement.

Putting the pieces together, we find that warlike races are perhaps *more* likely to achieve space travel. The ones with a history of victories will have the best technology, and will be most motivated to be at the very least wary of outsiders, if not openly hostile. This case can certainly be made for our own provincial example.

This hypothetical advanced civilization will be xenophobic, fearful of aliens. We've already seen that it's technologically possible to create interstellar starships, and it's also possible to engineer them to create duplicates of themselves, to speed up the time it takes to comb over the whole galaxy.

What happens when you take a paranoid species and give them the ability to build such spaceships?

Uh-oh.

The scenario plays out in the vignette at the start of this chapter. It makes a creepy kind of sense to me: any aliens that are that aggressive would want to wipe out potential enemies before they got sophisticated enough to pose a threat. The easy way to do it is to create space probes like the ones described, and use them to ruthlessly wipe out all life they find.

Death from the skies, indeed.

I've wrestled with this idea, wondering if it's possible. One potential saving grace is the same as before: exploring the whole galaxy in this way doesn't take long compared with the age of the galaxy itself. Therefore, according to the same logic above, it's likely that if such a xenophobic civilization were to evolve, *it would have been here by now.*

Yet we're still here. We know life has been around for billions of years. There have been the odd interruptions, but we've never been sterilized back down to the microscopic level. Like so many of the natural disasters we've seen, this puts a pretty good damper on the odds of being wiped out by nasty aliens. Simply put, if they were out there, we wouldn't be here.*

I honestly don't know if we're alone in the Universe; no one does. However, given the immensity of space, and the grand depth of time, it sure seems unlikely. And if we do get out there, it also seems unlikely we'll meet any nasty races like Klingons, Romulans, Vogons, Reavers, Daleks, or Kzinti. Natural disasters will still probably be our biggest worry.

But the galaxy is big, with room enough for lots of things. I may not know if we're alone, but I'd love the chance to find out.

---

* You might think that maybe they were here, 65 million years ago, and pushed the dinosaur-killer asteroid our way. But remember, they're advanced, smart, and without pity. A rock six miles across is pretty puny. They'd have dropped something a *lot* bigger on us, to make sure that in another few dozen million years, those little mammals crawling around the feet of the dinosaurs wouldn't evolve into a spacefaring threat.

CHAPTER 7

The Death of the Sun

THE PLANET IS FAIR-SIZED, CLEARLY BIG ENOUGH TO sustain a healthy atmosphere, though none is currently present. Given its distance from its parent star, it could easily have held liquid water on its surface too, once, in the far distant past. The outlines of continents are visible on its surface, though difficult to make out because of the lack of contrast. Were those deep, broad basins once ocean floors?

It's hard to tell now. The planet may have once been green, or even blue, but now it's all browns and grays and blacks. If any liquid water—or even water vapor—once existed there, it's long gone, evaporated a billion years before. Without an atmosphere there can be no liquid water.

The planet's star begins to peek over the planet's horizon. Swollen, distorted, nebulous, and very, very red, the star rises ponderously. It almost appears flat, it's so large. But after a few minutes the gently curved nature of the limb becomes more obvious, clarifying just how big the star is. An hour later it still hasn't fully risen, less than half of it showing above the horizon. It looms menacingly there, glaring like a bloody half-closed eye.

Finally, once the bottom limb clears the horizon, the cause of the planet's utter stillness and sterility is obvious. The star hangs over the landscape eating up a full 30 degrees of sky, as big as a dinner plate held at arm's length. The glowering eye of the star bears down on the planet's surface, which begins to heat up with the day. By midafternoon, the temperature is above the melting point of rock, and the surface of the dead planet begins to glow a soft red and liquefy once again. Mountains continue their slump, and continental shelves flow slowly, blurring into the dry ocean basins.

Finally, after hours of unleashing its crippling heat, the star sets, though its distended red glow lingers for hours. The rock begins to cool a bit, and by midnight is starting to resolidify. As the sky finally turns black, low wisps of rock vapor are illuminated from below by the still-molten lava shining through cracks in the ragged surface.

In a few hours, the cycle will start again. Every day, the star is marginally bigger, marginally brighter, marginally radiating more heat on the distressed planet. In a few more millennia the rocks will heat so much during the day that there won't be time for them to become solid again during the ever-briefer respites of night. The entire planet will become molten, erasing any possible hope of discovering its past history.

It's a shame. The planet's past is a rich one indeed, in its full and lively role as the third planet from the Sun. But the Sun has since started its long descent into death, and the past of Planet Earth will be lost forever.

SUNRISE, SUNSET

When you look at the Sun, it appears constant, stable, unchanging. But this is an illusion. Deep in its heart, an epic battle has been ongoing for billions of years, and will continue for billions more: the struggle between gravity and pressure. This war is fought with the weapons of contraction and expansion; more than anything else, the life of the Sun is defined by the balance between these two ancient foes.

Right now they are in rough equilibrium. The Sun's gravity is able to hold it together, counteracting its internal pressure that is trying to blow it up like a bomb. This uneasy balance has existed for eons, and will be maintained for a long time to come.

But not forever.

There is an old phrase, very old: *This, too, shall pass.* We use the Sun to measure the length of our days and years. These time scales are natural to us. But on much, much longer time scales the Sun itself will prove to be a clock that is running down.

The constancy of the Sun is an illusion. Like any other star, the Sun was born, and it will live out its life.

And someday, it will die.

SWIFTLY FLOW THE ~~DAYS~~ ~~MILLENNIA~~ EONS

The Sun's inevitable death isn't very pleasant to think about. But this eventuality *will* happen. An asteroid impact, a nearby supernova, a gamma-ray burst—these *might* happen. But they might not. The death of the Sun, however, is a rock-solid inevitability, and every second of every day brings us a tiny fraction closer to it. And one way or another, the end of the Sun also means the end of the Earth as we know it. It's even possible that it might literally mean the end of the Earth in point of fact; the planet itself may not survive the demise of its parent star. It certainly won't come out unscathed.

As described in chapter 3, the Sun cannot go supernova. While it is massive enough to fuse hydrogen into helium in its core as it does now, and will go on to fuse helium into carbon and oxygen, it just doesn't have what it takes to continue the fusion chain from there. It will never make iron in its core, and so it won't undergo the core collapse needed to power a supernova explosion.

So it will go out with a whimper and not a bang. But a whimper on the scale of a star is still a colossal event on a human scale.

Chapter 3 also gives a brief overview of what will happen to a star when it runs out of hydrogen in its core and begins to fuse helium. Some details not central to the discussion of how a star becomes a supernova were left out, but these details become critical when talking about the aging Sun. Perhaps you've heard that the Sun will one day turn into a red giant and engulf the inner planets . . . but saying that is like saying, "*Star Wars* is a movie about a kid who finds out he's cooler than he thought and winds up saving the day." The fun is in the details!

The Sun's timeline is defined by the laws of physics, and these are laws that will not be defied. They play out over much longer spans of time than we've covered so far, billions of years. Time keeps going whether we want it to or not, and we will see these stages of the Sun's life . . . and we'll see its inevitable death.

THE SUN AS A NORMAL STAR

Age of the Sun: 4.6 billion years (Now + 0 years)

Right now, we can consider the Sun to be roughly middle-aged: it's about 4.6 billion years old, and will live as a normal star for perhaps another 5 or 6 billion years.* It's currently steadily fusing hydrogen into helium in its core. That helium, over time, settles into the center of the Sun. The conditions there make hell look like the Antarctic: the temperature is 27 *million* degrees Fahrenheit, and the pressure is something like 250 *billion* times the atmospheric pressure at the surface of the Earth.

But even at these extreme conditions helium won't fuse into carbon and oxygen. The inert helium builds up in the very center of the core. It may seem odd, but the matter in the core still behaves like a gas, and obeys the same physical laws as a normal gas. As helium accumulates, the density of the core increases, and that means the temperature increases as well—a compressed gas heats up. This heat must go somewhere, so it radiates away into the upper layers and eventually out of the Sun as light.

This has been an ongoing process ever since fusion first ignited in the Sun's core 4.6 billion years ago. This means that for all that time,

---

* In the following sections, the number of years in the future should be considered approximate, perhaps accurate to a hundred million years or so.

very slowly, the Sun's core has been heating up as the helium in the core accumulates and compresses. This energy propagates through the Sun and is emitted out through the surface, so as the core heats up the Sun itself has been getting more luminous. It's something like 40 percent brighter now than it was at the onset of nuclear fusion all those eons ago . . . and it will continue to brighten as more helium is dumped in its heart.

THE BRIGHTENING OF THE SUN

Age of the Sun: 5.7–8.1 billion years (Now + 1.1–3.5 billion years)

The Sun's increasing brightness is a problem. The Earth is the temperature it is today because it intercepts a small amount of the Sun's emitted energy. But extra energy from the Sun means extra *heat,* which in turn will warm up the Earth. Because of the current global warming concerns, scientists have extensively studied the effects of temperature increase on the Earth's environment. If the Earth's overall temperature were to increase even as little as 10 degrees Fahrenheit, the polar ice caps would melt, causing an enormous environmental catastrophe.

The details of the Earth's temperature dependence on the Sun's energy output are complicated, but the overall effect is that as the Sun brightens, the Earth warms.* In general, this brightening happens slowly enough that life on Earth can adapt to the change. But when the Sun is about 10 percent brighter than it is today, the Earth's temperature will have risen those critical 10 degrees. The environmental impact will be profound. When the ice caps melt, the coastal regions of every continent will flood. The equatorial regions will become too hot for

* You might expect that the Sun's temperature is all that affects the Earth, but its size is important too. A ball bearing as hot as the Sun, for example, wouldn't heat the Earth at all because it's so small. Other factors in the Earth's temperature include its distance from the Sun, its ability to shed heat (radiating it away at night), and even how rapidly it rotates. However, all of these factors can be accommodated mathematically to produce a model of the Earth's temperature.

comfortable living, though areas like Greenland and even Antarctica will become warm.*

But polar warming probably won't balance the loss of habitability of the lower latitudes, because the Earth's air will dry out. When molecules of air heat up, they move around more quickly. Lighter molecules jostle around faster than heavier ones, and as the air heats up they can actually move rapidly enough to escape the Earth altogether! That's why the Earth's atmosphere has almost no hydrogen and helium; they are so light that they were lost to space billions of years ago. Heavier molecules like water, N_2, and O_2 tend to stick around.

But as the air gets hotter because of the brighter Sun, even heavier molecules can be lost to space. Eventually, the atmosphere will be too warm to retain water vapor. It will escape into space, drying out the air and leaving the continents of the planet as parched deserts. This will have obvious repercussions on terrestrial life.

Assuming a steady increase in the Sun's luminosity with time, that will occur in about 1.1 billion years.

That's a long time, and, in an odd way, it's a little bit reassuring! It doesn't take us off the hook for any other *current* environmental factors that contribute to any change in the Earth's temperature, but if you take the long view it does take the pressure off.

But that time will inevitably come. And things get worse after that.

The Sun will continue to brighten; after another 2.4 billion years (3.5 billion years from now) its brightness will have increased by 40 percent over today. The Earth's temperature will rise so much that the *oceans will totally evaporate.* The planet's atmosphere will still be too warm to hold on to all that water vapor, which will escape into space. The *entire*

---

* If you crunch the numbers, the average temperature of the Earth today at its current distance from the Sun should be just about or below the freezing point of water. It's warmer on Earth, on average, because we have an atmosphere. The greenhouse effect keeps us nice and toasty . . . but, of course, too much of a good thing doesn't help.

surface of Earth will be bone dry.* Sediment at the bottom of the oceans will be exposed to the heat from the Sun. As the ocean floors bake, carbon dioxide locked into the sediment will be released. Atmospheric carbon dioxide is a greenhouse gas; it lets heat from the Sun in but doesn't let it out. The Earth will heat even more, and the amount of CO_2 released will create a thick soupy atmosphere. It's entirely likely that in a few billion years, Earth will look very much as Venus does today: tremendously hot, and blanketed in a dense atmosphere composed almost entirely of carbon dioxide.

However, even that thick air will be lost to space over millions and billions of years. By the time the Sun's evolution brings it to the next chapter in its life, kicking it into overdrive, the Earth will be barren rock, devoid of any trace of atmosphere. It will be utterly lifeless.†

For those of you clinging to hope,‡ there is some life that *might* survive this stage of the Earth's distant future. Deep in a gold mine in South Africa, scientists found a colony of microbes that live off chemicals found there. The chemicals themselves are created by the natural radioactivity of the rocks, so these bacteria don't need any sunlight to live, which in turn means they can survive very deep underground. So, while life on the surface of the Earth will all die off over the next 3.5 billion years, life itself will continue somewhere in the Earth. That's cold comfort, perhaps . . . but it must be said: this, too, shall pass. After another 2 billion years, the Sun will start to *really* put the hurt on Earth.

---

* Actually, the Earth will be *drier* than bone, which is roughly 15 percent water by volume.

† An interesting coincidence is that life has been around on Earth for 3.5 billion years (give or take), and will continue for another 3.5 billion. We're currently right smack in the middle of the Age of Life on Earth . . . and any problems we have now may simply be chalked up to Earth experiencing a midlife crisis.

‡ My suggestion: let go.

THE SUN AS A SUBGIANT

Age of the Sun: 10.9–11.6 billion years (Now + 6.3–7.0 billion years)

So in this distant future, the Earth is dead, cooked to sterility by the ever-brightening Sun. However, while the story of the Earth is pretty much over at this point, the Sun's biography is just starting to heat up.

Because hotter it will most certainly get. Eventually, roughly 11 billion years after its birth, and 6.3 billion years from now, there won't be any more hydrogen left in the core of the Sun to fuse. The core will become entirely helium, but it still won't be hot enough to fuse into carbon and oxygen. Sitting on top of the helium core is roughly half the Sun's mass, pushing down on it, *squeezing* it, and the only thing able to support the core is its own internal pressure. As the Sun's mass bears down, the helium core will shrink even further.* As before, it responds by heating up. And up, and *up*. Although there is no hydrogen left in the core, the surrounding layers are lousy with the stuff—it's just that, until now, the pressure and temperature outside the core weren't high enough to cause the hydrogen there to fuse.

But at some point, the contracting core will reach a high enough temperature that the hydrogen in a thin shell surrounding it will fuse. This will add to the heat being generated inside the Sun, so the outer layers will respond by expanding. When this occurs, 6.3 billion years hence, the Sun's diameter will increase by about 50 percent, and its brightness will more than double. Astronomers call stars like these *subgiants*. They're bigger and brighter than before, but as the name implies, there's more to come.

The Sun will be a subgiant for about 700 million years. Over that time, its brightness will stay relatively constant, but its size will increase, from 1.5 times its current size to about 2.3 times its present diameter.

---

* Some studies show that the core will shrink by about 100 feet per year or so, which is not a whole lot compared with the core's size of many hundreds of thousands of miles across.

The color of the Sun will shift as well. As described in chapter 3, there is more energy being radiated, but a whole lot more surface area for it to be radiated *from.* Each square inch of the Sun actually emits less energy than before; it's just that there are more of them now. The surface of the Sun cools a bit, dropping a few hundred degrees, and the color becomes more orange than it is today.

The overall effect on the Earth at this point is minor. Fried from the increased energy over billions of years, the slight cooling of the Sun now hardly even makes a dent in the Earth. Life (except maybe for those subterranean bugs) is long gone.

Still, time marches on.

THE SUN AS A RED GIANT

Age of the Sun: 11.6–12.233 billion years
(Now + 7.0–7.633 billion years)

While the Sun is a subgiant, the core is still contracting and heating up. In the meantime, the hydrogen fusion occurring in the thin shell around the core is adding helium to the core as well. After the Sun has been a subgiant for about 700 million years, when it is about 11.6 billion years old, the mass of helium in the core will reach a critical point: it will become *degenerate.* This bizarre state is ruled by the laws of quantum mechanics and occurs when matter is compressed into incredibly dense states. Once matter is degenerate, it no longer behaves like a normal gas. For one thing, if you add mass to it, it responds by shrinking, the opposite of what you'd expect.* Also, the added mass does not increase the pressure inside the degenerate gas (this is very important later) as you might expect. Instead, just the temperature goes up.

The core will continue to contract as more matter is dumped on it

---

* When you take a ball of clay and throw more clay on it, it gets bigger. If you take a ball of degenerate matter and throw more on it, weirdly, it gets smaller. Quantum mechanics, it cannot be said enough, is really freaky.

by the hydrogen fusion. The temperature continues to rise, but not the pressure. As before, this added heat gets dumped into the outer layers, but the difference now is that the core is degenerate and heating up *a lot,* and it's doing it relatively rapidly.

When the core becomes degenerate, the Sun will be about 2.3 times as wide as it is now. But as the degenerate core heats up, the outer layers will respond once again to this added heat by expanding. When this process is done, the Sun will bloat to an incredible 100 to 150 times its size today—about *100 million miles in diameter.* The temperature will drop by half, and its luminosity—the energy it emits per second—will increase to a fierce 2,400 times its present rate. For the next 600 million years, the Sun will glow like a ruddy, fiery beacon. It is a *red giant.*

The view from the Earth will be awesome. Right now, you can easily cover the Sun with your outstretched thumb. As the Sun evolves during its time as a red giant, it will eventually *span a third of the sky.* It's difficult to appreciate how big that is. Go grab a yardstick. Put your left hand on one end, and your right hand at the 24-inch mark. Now extend your arms all the way out. When the Sun is at its greatest extent as a red giant, *it will just fit between your two hands.* Its growth would be imperceptible on a yearly basis, but over time it will grow to that immense size, appearing to loom over you in the sky.

Of course, you'd be fried long before then.

As the Sun expands into a red giant, several curious things happen. For one, its spin will slow almost to a standstill. When it expands, the spin slows in the same way a skater can slow her spin by throwing her arms wide. The amount the spin changes is more or less proportional, so if the Sun expands by a factor of 100, its spin will *slow* by 100. It takes a month or so to spin once now, so when it's a red giant it will take 3,000 days to rotate: more than eight years.

The Sun also becomes more luminous, as we've seen, but another critical change with its increased size is that its surface gravity gets lower. The gravity felt on the surface of an object depends on the mass of that object and its radius. When the Sun expands into a red giant, the mass stays the same but the radius increases by 100 times. This means the surface gravity will *drop* by a factor of 10,000 times. Currently, the surface gravity of the Sun is about 28 times that of the Earth (so that, for example, I would weigh well over two tons on the surface of the Sun*).

But when the Sun bloats up into a red giant, its surface gravity will drop to less than 1 percent of the Earth's gravity. Any particle on the Sun's swollen surface will only be very tenuously held on by gravity.

At the same time, the Sun's luminosity increases by 2,400 times. Any square inch of Sun will be blasting out 2,400 times as much radiation as it did before the Sun swelled up. This light has momentum that it can transfer to a particle on the surface, giving it an upward kick. To a random atom of hydrogen on the surface of the red-giant Sun, it will be as if someone had shut off the gravity at the same time he had turned on a huge fan from below: particles on the surface will be literally lifted off and blown away.

This stream of particles, called the *stellar wind,* is similar to the solar wind, but far denser. In fact, the red-giant Sun will shed something like one ten-millionth of its mass every year, far, far more than it does now through the solar wind.†

This mass loss is so great that during the time it takes to swell out to red-giant status, the Sun will lose a significant fraction of its mass. Since its gravity depends on its mass, its gravity will also drop. The planets, feeling a lower gravity, will migrate outward; their orbits will become bigger as the Sun loses its grip on them.

It's a race! Will the Sun increase in size quickly enough to engulf the inner planets, or will they migrate away from the Sun in time to escape its fiery maw?

For Mercury, the outcome is clear: doom. At 36 million miles from

---

* Even more if my wife just made cookies.

† Currently, the Sun loses only about 10^{-14} (one one hundred-trillionth) of its mass every year. Obviously, this is an incredibly small number.

the Sun now, it is too far behind the other planets even when the starting gun goes off. After a few million years, the Sun will catch up and expand right past the planet. Mercury will literally be inside the Sun.

What happens to it then? Interestingly, the outer envelope of a red giant is almost a vacuum. The mass of the Sun is still roughly the same, but the volume increases hugely; when it becomes a red giant the Sun will have a million times the volume it does now, so its average density will drop by that amount. In reality the density in the outer layers is even less than that, because a lot of the mass of the star is stored in the core. In the end, the density is less than one one-thousandth of the density of the Earth's atmosphere, almost a laboratory vacuum.

But there is matter there, thin as it may be. Mercury orbits the Sun once every eighty-eight days, so as far as Mercury is concerned it'll be sweeping through stationary material. As it plows through this matter, what is essentially air resistance will slow its orbital motion in the same way a parachute slows down a skydiver. In just a few years, Mercury will slow so much that it will spiral into the center of the Sun, where the increasing density of matter will accelerate the tiny planet's orbital decay. If it doesn't vaporize first, Mercury will fall into the center of the Sun, where it will most certainly meet its doom.

Pfffsssssst!

Of course, if drag on the Sun's matter slows Mercury down, the reverse is true as well: Mercury will accelerate the particles in the Sun's outer layers. As Mercury slowly spirals into the Sun, it will speed up the Sun's spin. It won't be by much, just a percent or two. By the time the plunge into the heart of the star is over, the only indication that the solar system ever had a planet called Mercury will be a very slight increase in the Sun's spin.

What of Venus? As it happens, our knowledge of how the Sun will expand into a red giant is still a bit too uncertain to know if Venus will evade getting eaten or not. Some models show it escaping, while others show it suffering the same fate as its little brother. Even if it does manage to stay outside the Sun's greedily expanding surface, Venus is doomed. From just a few million miles away, *the Sun will fill Venus's sky.*

The surface of Venus is hot to start with—900 degrees Fahrenheit, thanks to its runaway greenhouse effect—but when the Sun looms so terribly over the Venusian surface, the temperature will scream upward to nearly that of the Sun itself. Venus's crust will melt and its atmosphere will be blown away.

The Earth may fare somewhat better. Some studies show the Earth's orbit expanding more quickly than the Sun, while others show us being consumed by the ever-growing star. Astronomers are still arguing over the details, which are important in this game of catch-me-if-you-can.* Depending on the details of how the Sun expands and how much mass it loses, the Earth will end up being about 1.4 times farther from the Sun than it is now. The Earth is currently 93 million miles from the Sun, so when the Sun stops expanding that distance will increase to 130 million miles.†

Even if we escape being engulfed, don't breathe a sigh of relief just yet: remember, the red-giant Sun is huge. It will fill a large fraction of the Earth's sky, radiating down on it at a temperature of over *5,000 degrees Fahrenheit.* The Earth's surface temperature at that point will be about 2,500 degrees, hot enough to melt nearly every metal and rock on its surface. Even before the Sun swelled up the Earth was quite dead, its oceans having boiled off and the atmosphere ripped away. But during the Sun's red-giant phase, the crust of the Earth will melt as well, and that, pretty much, will be that. While it's not literally the end of the world, it's certainly the end of the world as we know it.

We still have room for one more "however," however. While the Earth will be totally stewed, it's not the only usable planet in the solar system. Mars too will move out from the Sun, but, unfortunately, will also be too hot for life. Even Jupiter's moons will warm up too much to

---

* Many older books on astronomy say that the Earth will definitely be swallowed up by the Sun when it becomes a red giant, but those works don't account for the Sun's mass loss through its supersolar wind and the subsequent increase in the orbital diameters of the planets.

† Remember, that's the distance from the *center* of the Sun. The surface will be 50 million miles or so closer.

sustain us (the average temperature will be something over 500 degrees, hotter than your kitchen oven when you bake cookies). Jupiter's moon Europa is an icy body, and thought to have liquid water under the surface. When the Sun expands into a red giant, Europa might entirely vaporize.

It's possible that no existing place in the solar system will be cool enough to support life as we know it. Even the distant moons of Uranus and Neptune will be too warm. You'd have to be about 4.5 billion miles away from the Sun to get temperatures near where they are on Earth today. Of all the places in the solar system, in six billion years only the (currently) icy bodies slowly orbiting the Sun well past Pluto's orbit may be cool enough for us. They would melt all the way through, becoming essentially giant drops of water a hundred or so miles across, with a red, swollen Sun glaring down on them. It's known that currently these objects are loaded with organic chemicals. When those icy bodies warm up, all sorts of interesting things could happen to those chemicals. The bodies will stay liquid for hundreds of millions of years while the Sun remains a red giant, which begs the question:

What life might evolve under those circumstances?

ASIDE: DAVID AND GOLIATH

At this point in the life of the solar system, things don't look so good for the home team. The Sun is a swollen, distended blob, it's eaten one planet, fried three others, vaporized a retinue of moons, and generally made things uncomfortable for almost everyone else.

But what are you gonna do?

Actually, that's an excellent question. So far, this story has unfolded in this manner because we've let it. That is, if we sit back and watch, this is the way it will play out.

But it doesn't have to be that way.

For example, it will take several hundred million years for the Sun to go from subgiant to giant. During that time, the temperature on Earth will be unbearable. And once the Sun does go all the way to giant, even

Mars won't look so good. But a billion years is a long, long time, and during that time Mars may be the place to be.

It's smaller than the Earth, and has very little atmosphere. We can't do much about its small size, but we can bring it air . . . by dropping bombs on it. Bombs in the form of comets.

Comets are large chunks of rock and ice, and some, in the distant outer solar system, are quite large, hundreds of miles across. They move so slowly in that far realm that it wouldn't take much of a kick to drop a few into the inner solar system. Attaching a small rocket to one would do the trick. Letting smaller pieces hit Mars, one at a time, could easily bring enough water to fill ponds, lakes, and eventually oceans. Careful manipulation of its atmosphere, using genetically engineered plants and chemical processing, could encourage the development of breathable air. Some people estimate it might only take a century or two.

This type of practice, making a planet more Earthlike, is called *terraforming*. It's a staple of science fiction, but it's based on fact; the physics, chemistry, biology, and other fields of science involved are generally well understood. The devil's in the details of course, but we have plenty of time to work them out. I'm guessing that in that dim future, billions of years down the road, a century or two here and there will hardly matter.

In fact, we'll have the technology to start work like this on Mars much sooner than six billion years from now; realistically it could start early in the next century. Which raises the question: in six billion years, won't we have terraformed all the planets? Perhaps. With an ever-burgeoning population, future humans will look across the gulf of space at all that real estate with envious eyes, and slowly and surely, as H. G. Wells once wrote, they will draw their plans against them.

Still—and stop me if you've heard this before—there's no place like home. The Earth is a pretty good place, and we've spent a lot of time evolving here to make ourselves fit in. Is there no hope for our home planet?

Actually, yes, there is, and the solution is simple: we just need to move it farther from the Sun.

How hard can *that* be?

Okay, in practice, pretty hard. The main problem is that the Earth is a big, massive object, so moving it takes a *lot* of energy.* To move the Earth out to where the temperature will be more hospitable (around Saturn's current orbit) takes roughly the same amount of energy that the entire Sun currently emits in a solid year. That's the equivalent of exploding 200 quadrillion one-megaton nuclear bombs.

There might be some environmental effects from that.

There are alternatives. We could strap a few million rockets nose-down onto the Earth and fire them off, but it's hard to know where we'd get enough fuel for them. Plus, the Earth's rotation and revolution around the Sun would complicate things (we can assume, however, that by the time we need to do this our technology will be up to the task).

But there's a better way, the environmental impact of which—if we're careful—is essentially zero.

When we send probes to the outer planets, we can give them a boost in speed by "borrowing" (really, stealing) energy from the orbital motions of other planets they pass on the way. This is the so-called *slingshot effect.* If we want the probe to speed up, we send it on a path so that it comes in from "behind" a planet, catching up to it as the planet moves in its orbit around the Sun. As the probe passes the planet, it picks up some energy from the planet's orbital motion, which increases the probe's speed. The planet *loses* the same amount of energy, and slows down a bit in its orbit. Since a planet is typically a lot more massive than a space probe, it slows down very little, an immeasurable amount really, while the probe can gain quite a bit of speed. This means we can send probes to the outer planets without having to carry vast amounts of fuel.

Taking energy away from a planet will drop the planet ever so slightly

---

* In reality, a bigger problem might be that all the plants on Earth have evolved to make oxygen using the color of sunlight we have now. A much redder Sun may be a much larger headache for our descendants than such a trivial thing as moving the Earth.

closer to the Sun. But, if we do this in reverse—send the probe in "ahead" of the planet—then the probe *loses* energy, giving it up to the planet. The probe slows and drops closer to the Sun (useful for getting probes to the inner planets, such as Mercury) while the planet gains energy, moving *outward* from the Sun.

If we want to move the Earth farther out from the ever-increasingly sizzling Sun, this is an excellent way to do it. Instead of space probes, we can use asteroids, which are much more massive. That means the energy exchange is greater, requiring fewer slingshots. Moving asteroids isn't all that difficult; in that case strapping a rocket onto one would work pretty well. By aiming it just so, the asteroid could give some of its orbital energy to the Earth, moving the Earth just a teeny bit outward from the Sun. Lather, rinse, repeat . . . a million times.

This scenario has been studied by the astronomers Donald Korycansky, Greg Laughlin, and Fred Adams, and they found that by using a large but typical asteroid, such a maneuver could feasibly move the Earth slowly out to a safe distance from the Sun.

Here's how you do it. Start with a large rock about 60 miles across that is well out in the suburbs of the solar system. Change its orbit using a rocket or some other method so that it drops into the inner solar system. Aim it (here a rocket would be useful for fine-tuning) so that it passes in front of the Earth, missing us by about 6,000 miles.* The exact amount of energy transfer depends on a lot of factors, such as the angle of the incoming rock, how close it passes, and so on, but in general a single passage of a rock this size would add about ten miles to the Earth's orbital radius.

That's not much, of course, but at first it doesn't need to be much. Small steps are okay; we have plenty of time!

From this point we have two options. We could, for example, wait a

---

* This is actually a terrifyingly close encounter. At closest approach the asteroid would be nearly as large in the sky as the full Moon—features on the surface would be easily visible to the naked eye—and be moving so rapidly that it would cross the sky in just a few minutes.

few thousand years, find a second asteroid, and have another pass. But this is wasteful; for one thing there aren't enough asteroids of this size in the solar system to do the trick. We'll run out while the Earth is still too close to the Sun.

A second option is better: instead of simply throwing away the first asteroid, we recycle it. A little preplanning and care can save the day. Instead of letting the asteroid go away, we time the passage so that as it heads back out into deep space, it passes by either Jupiter or Saturn. This time, though, it passes *behind* the planet, gaining energy. Then the orbit can be adjusted again (using the onboard rocket; if it uses solar energy we even get our fuel for free) to pass by the Earth another time. If we do this, the asteroid becomes a sort of interplanetary orbital energy messenger, taking energy from Jupiter or Saturn and delivering it to Earth.

As Earth moves out, Jupiter will move *in*—remember, we're stealing its energy—but Jupiter is so much more massive than the Earth (300 times, in fact) that it migrates far less than the Earth does. Moving the Earth outward far enough to keep it temperate while the Sun is in its subgiant phase (about 50 million miles or so) will require Jupiter to move only a few million miles inward (it is currently about 400 million miles from the Sun).

This will pose a problem when the Sun evolves into a red giant, however. The Earth will have to move out past where Jupiter is now. We could still steal from Jupiter's orbital energy to do this, but once the two planets get near each other Jupiter's gravity starts to affect the Earth directly.* Any encounter like that between the largest of the solar system's planets and us is bound to have an unfortunate outcome: the most likely scenario is that the Earth gets ejected from the solar system altogether (see chapter 5).

It's possible we could use a second set of asteroids to move Jupiter

* When the Sun loses mass, Jupiter will migrate outward as well, but we'll also be stealing its energy, which moves it *inward*, so it's hard to say just where it will end up.

out farther from the Sun as well by stealing energy from Saturn, Uranus, and Neptune. At this point, though, the math gets pretty complicated, and results are difficult to pin down.

However, we have a few billion years to work out *exactly* how we'll play musical planets. We've probably figured it out well enough for now. It's a viable system, and one our descendants may very well have to employ.

THE HELIUM FLASH AND CORE HELIUM FUSION

Age of the Sun: 12.233–12.345 billion years
(Now + 7.633–7.745 billion years)

So now the calendar reads 7,633,000,000 AD (give or take a millennium), the Sun is a huge red giant, just reaching its maximum extent to a diameter of over 100 million miles, Mercury is gone, Venus may still be around but suffering, Earth is still here but possibly orbiting much farther out than it did when the Sun was middle-aged, and Pluto is a prime condo spot (complete with planet-spanning swimming pool). A time traveler from the twenty-first century would hardly recognize her neighborhood.

But we're not done, not by quite a bit. The Sun won't stay a red giant forever. And, as usual, the key to what's happening lies deep in its heart.

The core of the Sun is now pure helium, and contracting. It's degenerate, and heating up. The hydrogen around it is fusing into helium in a thin shell, adding more ash to the core. Remember too that since the core is degenerate, its pressure doesn't change as mass is added. The temperature keeps going up, though.

At some point, something like 600 million years after beginning its transformation into a red giant, the core reaches a temperature of 100 million degrees Fahrenheit. Then all hell breaks loose. Well, to be more accurate, all hell is *released,* but it doesn't break loose.

At that temperature, helium can fuse into carbon. Now, if the core were just a normal ball of gas heated to that temperature, the helium would fuse, heat would be released, and the gas would expand, adjusting its internal pressure to accommodate the extra heat—this is essentially what the core and outer layers of the Sun have been doing for billions of years, playing temperature, gravity, and pressure against one another.

But the core *isn't* normal. It's *degenerate.* It can't adjust its pressure. So as the temperature increases, it cannot increase its size to compensate. Somewhere, deep in the core, the temperature reaches that critical point, and fusion of helium into carbon begins. This releases energy, which raises the core's temperature.

This is bad. The fusion rate for helium is *ridiculously* sensitive to temperature. A slight increase in temperature and the fusion rate screams up, raising the temperature even more, again increasing the fusion rate. Within literally seconds this vicious circle runs away, and the inside of the Sun's helium core explodes like a bomb.

The energy release is difficult to exaggerate: it's colossal, epic, titanic. In that one brief moment, called the *helium flash,* the core of the Sun releases as much energy as *all the rest of the stars in the galaxy combined.* It may actually release *100 billion* times the Sun's normal output, all in a few seconds.

You'd think this would tear the star apart in a supernova, but in fact, it doesn't. It does a funny thing: because this is all happening deep inside the core, the matter itself absorbs all the released energy. This infusion of energy is enough to overcome the degeneracy of the core, which suddenly becomes normal matter once again. It's under tremendous pressure, to be sure, but it's no longer held in the sway of that weird quantum mechanical state. Once the degeneracy is released, the runaway fusion flash is dampened, and everything settles down into a nice steady state.

With that huge explosion safely absorbed, and the core back to being a regular old gas, helium fusion can proceed at a more leisurely pace. So now we have a core of helium fusing into carbon (there are also some minor avenues of fusion that are producing oxygen and neon as well), releasing heat. Outside this is still a thin shell of hydrogen fusion, and surrounding that, a hundred million miles deep, are the outer layers of the star.

Ironically, however, the amount of energy being generated in the core due to helium fusion is now *less* than was emitted when the core was degenerate and shrinking. This means less heat is being transported into those deep outer layers, which were before being supported by that extra heat. Once the core cools off, the outer layers respond by shrinking back down. On a relatively short time scale (about a million years), just as the red giant is reaching its maximum possible size, the legs are kicked out from under it. The Sun shrinks.

When it settles down, the Sun has become considerably less bright, emitting now only about 20 to 50 times as much energy as it did when it was young, only a few percent of what it did at its peak as a red giant. It's still bigger than it was when it was a normal star, but far smaller than a red giant: it's now about 10 times its original size, 8 million to 10 million miles across. It's slightly hotter now, radiating away at about 8,000 degrees Fahrenheit, still cooler than its temperature today as well. It's a lovely orange in color.

Since it's smaller, the Sun's surface gravity increases (even though it lost some mass as a red giant). Particles on the surface are held on more strongly. Moreover, the luminosity has dropped, so the particles feel less of a pressure to blow off the surface. The stellar wind decreases drastically.

So now the Sun is stable once again. It'll remain this way, a helium-fusing giant, for over a hundred million years.

The Earth, however, is once again in trouble. After all our effort to move it a billion miles out, we suddenly find that the Sun is much smaller and giving off less energy. Temperatures plummet. Those far distant descendants of ours will have to move the Earth *back* toward the Sun. No problem—they can do the reverse of what they did to move it out. They have a lot less time to do it, though: they had billions of years to migrate it outward, but now they'll have to drop it inward in only a million years or so. They can use bigger objects (Jupiter has lots of moons it doesn't need, for example) to increase the rate of energy transfer.

Or, who knows? *It's more than seven billion years in the future.* Maybe they'll just snap their fingers and the Earth will tunnel through space-time and reappear where they need it.

Let's hope it's that easy. Let's also hope they're patient and not easily irritated, because in a few dozen million years, we're going to start all over again.

In the Sun's core, carbon and oxygen are building up. The core is too cool to fuse them, so they are inert, like helium was before them, accumulating like ash in a fireplace. So the scenario is familiar: the core starts to contract, and the Sun slowly starts to heat up *again.* Over the next 20 million years it slowly starts to brighten and swell. After having moved the Earth out and back in again, we'll be forced to migrate our planet away from the Sun once more. The outer layers of the Sun will reach an extent of 20 million miles or so before the next catastrophe occurs.

HELIUM EXHAUSTION

Age of the Sun: 12.345–12.365 billion years
(Now + 7.745–7.765 billion years)

That happens when the helium in the core runs out. The carbon/oxygen core starts to contract, just as the helium core did before it, and the results are similar: the Sun, for a second time, becomes a red giant. This time, though, the onset is much faster. Carbon and oxygen have different physical properties from helium's, and the core contraction is more rapid. Instead of taking 600 million years to expand, it takes only 20 million.

The Sun expands drastically again, achieving a diameter of well over 150 million miles. Its sudden expansion will make our human celestial engineers pull their hair out.* Probably, at this point, it's a good idea to abandon the solar system and look for lodging elsewhere.

It's all for the best, perhaps. The view from far away will be spectacular, as we'll see in a moment.

The Sun will be more luminous in its second regime as a red giant than it was the first time. It will blast out energy at 3,000 times the rate it does now, and the stellar wind will be back with a vengeance. It lost 28 percent or so of its original mass the first go-round; this time it loses more than 60 percent of what is left. With or without our help, the planets will once again migrate outward as the Sun hemorrhages away its material, with Venus and Earth possibly moving quickly enough to avoid being consumed once again. And if they escape they'll *still* get roasted once again by the swollen, luminous Sun.

This is a grueling series of events for the solar system. Yet, amazingly, things are about to get worse.

Deep in the Sun, the carbon/oxygen core gets so dense it becomes degenerate. Helium fusion starts up in a thin, slightly degenerate shell outside of it, and hydrogen fusion continues in a shell outside of *that.* The problem is, thin-shell helium fusion is wildly dependent on temperature, even more so than before. Any slight increase in temperature causes the fusion rate to increase madly.† As more heat is generated, the rate goes up, which generates more heat—well, we've seen this before. The thin helium shell can flash again, releasing huge quantities of energy. This time, though, the outer layers of the Sun won't have time to expand slowly and accommodate the extra energy. The rate of energy being dumped into them simply overwhelms them. The Sun convulses, literally, and ejects a vast amount of material over the course of just a few years—not millions of years, mind you, just plain old *years.*

---

* Yes, assuming they have any. Or hands. Or heads.

† Helium fusion under these circumstances has a rate that scales as the temperature to—hold on to your hat—the *40th* power. This means a teeny-tiny increase in temperature causes the rate of fusion to accelerate insanely; a 20 percent rise in temperature increases the helium fusion rate by 1,500 times!

After the flash of energy, the helium shell cools down for perhaps 100,000 years, but then the situation builds again. A second flash occurs, and a second envelope ejection. Then, again after 100,000 years, a third, and then a fourth, most likely final, flash and ejection. During these episodic convulsions, the Sun swells for a third time, this time expanding to as much as 200 million miles across, enough to reach the Earth's original orbit.

Even at its increased distance, the Earth won't fare well during these eruptions. Its surface temperature will rise to well over 2,000 degrees as the swollen Sun heats it, then drop again after each pulse fades. Also, these pulses will slam the Earth with quadrillions of tons of matter moving at several miles per second. This won't add much to the total mass of the Earth (which is thousands of times more massive than the material accumulated), but the impact of that much material, even spread out over hundreds of millennia, will severely batter the already war-torn Earth. The total impact energy is equal to the detonation of *trillions* of nuclear weapons, or the same as detonating a one-megaton bomb every second for a million years.

Even in death, the scale of destruction wrought by the Sun is awesome.

Every time the Sun erupts, it loses more mass. By the fourth epic heave, the last bits of the outer envelope will be shed. The majority of the Sun's original mass will be lost to space, revealing just the degenerate carbon/oxygen core surrounded by a thin shell of very hot helium. The core has contracted to just a few thousand miles across, about the size of the Earth (assuming our planet still exists). It will have about half the mass of the original Sun, so it is phenomenally dense. It's also still quite hot; it will radiate at a temperature of as much as 200,000 degrees Fahrenheit, and will shine at thousands of times the luminosity of the present-day Sun.

It has become a white dwarf. To someone standing on the surface of the blasted and quite dead Earth, the Sun would only be a point of light, eye-achingly bright, brighter than the full Moon is now. But it will be only a pale, dim shadow of its former glory.

We're pretty much at the end of the line here. No more fusion, no more source of energy. After a lifetime of over 12 billion years and a dramatic saga of expansion, contraction, and eruption the Sun is, effectively, dead.

THE SUN AS A WHITE DWARF

Age of the Sun: 12.365 billion years (Now + 7.765 billion years)

However, in death there can be beauty.

The gas ejected from the Sun in its final days will be expanding rapidly. The distribution, the overall shape, of the gas will depend on many factors. In general, the Sun will emit the gas in great spherical shells like cosmic soap bubbles, with a dense edge and more tenuous inner region. However, there can be circumstances where the gas can be shaped, molded. As it expands it might hit gas that floats between the stars (what astronomers call the *interstellar medium*). If the Sun was spinning rapidly enough when the gas was emitted (absorbing Mercury and Venus might just be able to speed it up, since the planets would dump their angular momentum into it), the shells might be flattened by centrifugal force, shaped more like a cheese wheel or a basketball with someone sitting on it. Other physical conditions can cause clumps in the gas, or bright regions, or rings.

The white-dwarf Sun, sitting at the center of this expanding gas, may be hot enough to flood surrounding space with ultraviolet light. This would ionize the gas, causing it to glow.

If our descendants have fled to another star, what a sight they will see! Looking back on our Sun, they may see the gas glowing like a perfectly circular thin ring. The ring will glow mostly green because of the oxygen atoms in it; other elements contribute different colors, but the green glow is generally the strongest because of the way the atoms of oxygen emit light. Through a small telescope the greenish disk will resemble the planet they once lived on; astronomers today call these objects *planetary nebulae* for that reason.

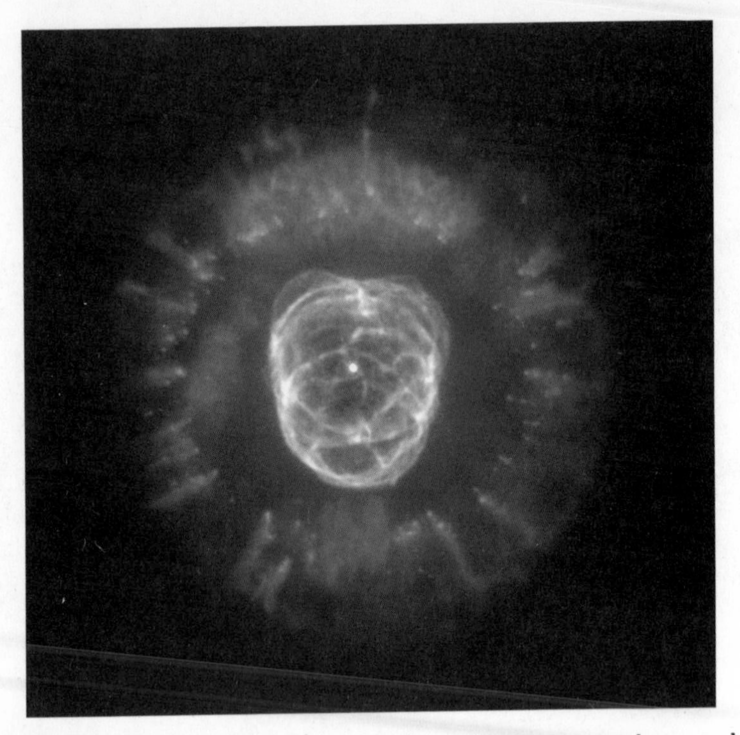

Planetary nebulae form when a star like the Sun dies, ejects its outer layers, and ionizes them. The gas glows, forming eerie shapes. This Hubble picture is of the famous Eskimo Nebula, which indeed looks like a parka-wearing Inuit in ground-based telescope images.

NASA, ESA, ANDREW FRUCHTER (STSCI), AND THE ERO TEAM (STSCI + ST-ECF)

Planetary nebulae are among the most beautiful objects in the sky. How will those distant humans feel, looking back at their solar system? Will they feel any better at all, knowing that civilizations across the galaxy will be able to view our Sun and see its final attempt at glory? Or is it silly to try to even guess what humans, if there are any, would be feeling more than seven billion years from now?*

---

* I hate to say it, but some calculations indicate that the white-dwarf Sun won't be bright enough to ionize the expanding gas before the material disperses into interstellar space. It's likely that when the Sun has this final fling, it will be too dark to see.

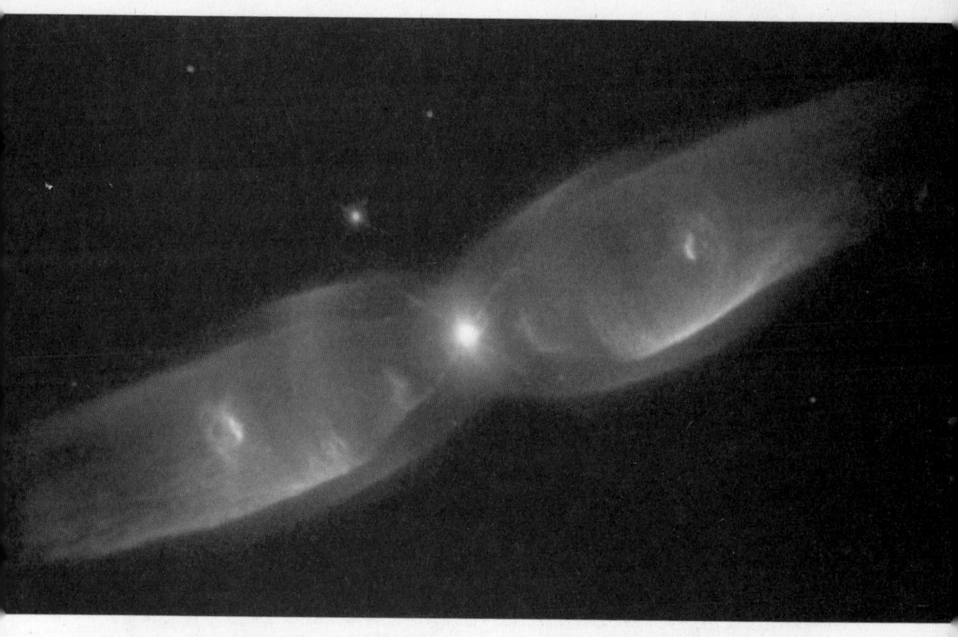

M2-9 is another planetary nebula, but has a more elongated shape. This may be due to the presence of a binary companion to the dying star; if the red-giant star swallows up the companion, the wind it emits as it dies can be sculpted into odd shapes.

BRUCE BALICK (UNIVERSITY OF WASHINGTON), VINCENT ICKE (LEIDEN UNIVERSITY, NETHERLANDS), GARRELT MELLEMA (STOCKHOLM UNIVERSITY), AND NASA/ESA

Finally, a few thousand years later, the gas will have expanded and thinned, and the white dwarf cooled. There won't be enough UV light to illuminate and fluoresce the gas, and not enough gas to absorb it anyway. The expanding material that once warmed and cheered our planet will merge with and become indistinguishable from the gas that exists between stars. The white dwarf will continue to radiate away its heat, inevitably cooling, dropping in color from white to blue to yellow to orange to red, and then it will slide to infrared, and invisibility, after a few more million years.

Whatever is left of the solar system will continue to orbit the now-black dwarf. The planets will cool along with their star, eventually freezing solid, and after a few billion more years will be as dark and cold and empty as space itself.

THE LONG DESCENT INTO NIGHT

Can planets survive such a devastating series of events? Actually, the answer is yes. Depending on what you mean by "survive."

First, well over two hundred planets have been found orbiting other stars, and a dozen or so of these have been found orbiting red giants. These planets are probably something like ours: they formed along with their star billions of years ago, and have managed to make it through at least one episode of red-giant expansion. We don't know if those stars swallowed up any inner planets or not, but at least some planets were outside the zone of envelopment. The planets may have started rather close to their stars, though, and migrated outward; one, for example, is not much farther away from its star than the Earth is from the Sun. The surface temperature of that planet is probably about 900 degrees Fahrenheit.*

Amazingly, evidence has recently turned up for planets orbiting white dwarfs. A strong signal of the element magnesium was found coming from a white dwarf, far more than could possibly be generated by the star itself. The amount of magnesium found indicates that quite recently an asteroid must have come too close to the star, been disrupted by the dwarf's strong tides, and formed a disk of matter around the star. The shape of the disk itself means that the asteroid may have been in a much larger orbit, only to be dislodged by a planet or some other massive body. What all this means is that planets and even asteroids may survive not only a star's evolution into a red giant but also all the subsequent catastrophic stages. The white dwarf itself is fairly massive, about 0.8 times the Sun's mass. When the Sun becomes a white

---

* The planet in question, orbiting the red giant HD 17092, has a mass more than four times that of Jupiter, so it's almost certainly a gas giant with no solid surface. Therefore, to be pedantic, the temperature is 900 degrees at the top of its cloud layer.

dwarf it will have less than half its original mass, indicating that this star must have started off *more* massive than the Sun, which means in turn its later stages in life would have been even more extreme than the scenario outlined above.

Yet, even there, it appears that planets were able to hold it together during those hundreds of millions of years of stellar torture. Of course, by the end of this period of repeated roasting and freezing the planets would be burned-out sterilized hulks. There must not have been any civilizations there capable of manipulating planetary orbits, or able to foresee the future well enough to plan for this eventuality.*

But we're smart, we humans. I can hope that when the time comes—*and it will*—we'll be able to do something about it. And I really do hope we—or something resembling us—will be around. The death of the Sun will be sad to behold . . . but the beauty will be breathtaking.

---

* This may seem depressing to some, but it's a relief to me: I'm not sure I want celestial neighbors capable of engineering on that scale.

CHAPTER 8

Bright Lights, Big Galaxy

YOU WALK OUTSIDE ON AN EARLY WINTER'S EVE AND cast your gaze upward. The constellations in the north and east reflect the tale of heroic Perseus, sent by King Cepheus and Queen Cassiopeia to rescue the maiden Andromeda from the sea serpent Cetus. You may chuckle, thinking that two curved lines of stars hardly bring to mind the vision of a young girl chained to a rock waiting to be eaten by a monster, but as you look carefully your eye is caught by a glimpse of something fuzzy just off Andromeda's side. It's hard to see, but it's definitely there: a slightly elongated patch of light.

It seems suspended there, small and unassuming, hanging for all eternity in the sky. But, like so much about the night sky, that's an illusion. You are seeing the great Andromeda galaxy, a massive spiral galaxy similar to our own Milky Way. And it's headed our way.

If you could speed the clock up a few trillion times, such that millions of years appeared to elapse in mere seconds, you would see the Andromeda galaxy swell and grow before your eyes. Every passing moment would see it getting larger, until a few minutes later it looks like the whole sky is about to fall on you. You see stars around you suddenly wrenched up and away, forming a long thin line stretched into space, like a tendril reaching out to the other galaxy. The rest of the sky is oddly empty of stars, the Sun too becoming a part of a stellar stream stretching countless light-years, reaching into the space between the galaxies where stars are rare.

Suddenly, the Andromeda galaxy has flown by, shrinking in the distance somewhat, having literally passed through our galaxy like a ghost through a wall. However, over the next few accelerated minutes—actually, millions upon millions of years—you see it slow, stop, and then head back your way. Flash! It fills the sky in another pass, and then once again has moved away. But this time it doesn't get as far. Andromeda swells one last time. Over your head you see the bright core of the galaxy merge with the core of our own. Above you hangs a vast cloudy ball, the remnants of the once mighty pair of galaxies, merged to form a single, larger galaxy.

Within minutes, the sky settles down. Everything now looks calm. You sigh with relief, glad that you have survived this cosmic encounter. What you don't know is that a beam of matter and energy is headed your way from the heart of the new galaxy, and when it touches down on the Earth, the chaos of the merger will seem as bucolic as a peaceful springtime meadow.

THERE'S NO PLACE LIKE HOME

Have you ever heard that a galaxy is like a city? A city has a downtown section, suburbs farther out, pockets of congestion, regions where there's not much to see, and, of course, the occasional rough neighborhood. Galaxies are like that too. They have their regions of high and low activity, places that are exciting, places that are a bit duller. We even say they have a population—but instead of people, a galaxy is populated by stars, gas, and dust.

And, like a city, of course, there are places you really don't want to go.

Yeah, you can see where this is headed.

If you live far from civilization, in a place with dark skies, then you have certainly stepped outside on a clear, moonless night. At first, when you do so, you may only see a few stars in the sky as your eyes slowly adjust to the gloam. But over time, as your pupils dilate and your eye automatically coats your retina with a light-sensitive protein that increases sensitivity, fainter stars will slowly become noticeable. The sky will become spangled with stars, thousands of them gently twinkling down upon you.

Along with the stars, you may see a faint glowing band across the sky. It almost looks like smoke, or a jet contrail. That swath of mist is called the Milky Way, named because it looks like a river of spilled milk across the sky. It has been known for thousands of years, since humans first noticed the night sky. In the early 1600s, Galileo turned his newly fashioned telescope to the Milky Way and was shocked to see that it was not a cloud, but was in fact made up of thousands of stars, all too faint to be seen individually.

This was the first clue that we live in a *galaxy.** A clue to its shape may come with a moment's inspection of the brightest stars in the sky, revealing that they seem to stick near this milky band. Away from the band, stars are fainter and fewer in number.

In the eighteenth century, the great astronomers William and Caroline Herschel took this idea further: using a telescope, they counted up stars in different directions in the sky to try to determine the shape of the galaxy.

The idea goes something like this: imagine you are standing in a field. As the sky darkens, you are surrounded by fireflies. If you're in the middle of the field you'll see the same number of fireflies in every direction you look. But if you're near the edge of the field, you'll see fewer fireflies toward the edge than if you look across the field to the far side.† The farther you can see, the more fireflies you see.

So it goes with stars. The Herschels reasoned that if they saw more stars in one direction than another, then that means the galaxy must be

---

* The word *galaxy* comes from the Greek word *galaxias*, meaning milk, a reference to the Milky Way Galaxy. There is some confusion over the term *Milky Way;* sometimes it means the galaxy itself, and sometimes the milky stream of unresolved stars you can see from your backyard. Usually it's clear in context.

† Yes, this is the same analogy used for GRBs in chapter 4. Glad you noticed! The principle is the same, so I recycled it.

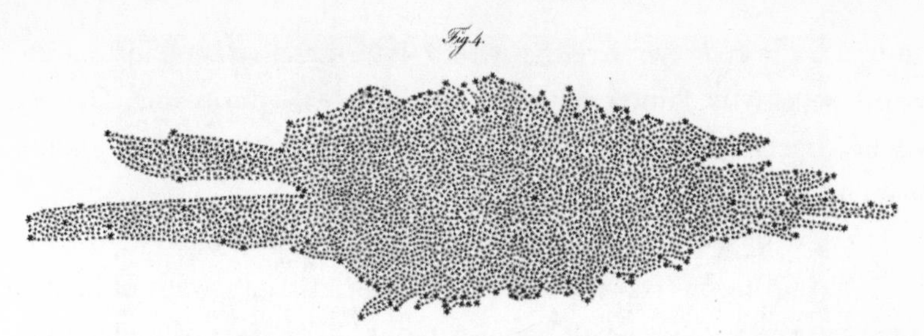

The shape of the galaxy as determined by William and Caroline Herschel in 1785. The Sun is near the center of a "grindstone"-shaped galaxy.

WILLIAM HERSCHEL, 'ON THE CONSTRUCTION OF THE HEAVENS,' PUBLISHED IN *PHILOSOPHICAL TRANSACTIONS OF THE ROYAL SOCIETY OF LONDON*, VOL. 75 (1785), PP. 213–66. IMAGE COURTESY RICHARD POGGE.

longer in that direction. They found the galaxy to be a flattened cigar shape with the Sun very near the center.

This method was repeated in 1906 by another great astronomer, Jacobus Kapteyn, using photographs instead of eyeball observations. He started his star-counting exercise, and in the end determined that the galaxy is roughly cigar-shaped, about 45,000 light-years across, with the Sun very near the center.

This idea was brilliant, but unfortunately it doesn't work well in practice. In both attempts, the numbers, and position for the Sun, were way off. Why?

They didn't know about dust. If cities have pollution, galaxies have dust.

It's not the kind of dust you find clumping underneath your sofa or dimming your TV screen. Galactic dust is actually composed of complex carbon molecules. Stars create this dust in their stellar winds, while simultaneously blowing it into space.* In sufficient quantities dust is opaque and blocks starlight.

This put the kibosh on the star-count method. Imagine now that instead of a field full of bugs, you are in a large room filled with smoke. The

---

* There are other sources of dust as well, including supernovae, but red-giant stars near the ends of their lives are the primary source.

The Milky Way Galaxy is a grand design spiral, epically beautiful. This artist's interpretation is based on actual observations, and accurately depicts the shape of our galaxy. The Sun is roughly halfway out from the center. Although the disk spans 100,000 light-years, it is only about 1,000 light-years thick, so viewed edge-on our galaxy would look very flat, like a pancake with a bulge in the middle.

NASA/JPL-CALTECH/R. HURT (SSC)

smoke is so thick you can only see a few feet in each direction (think billiards parlor). Since your vision is so limited, you have no idea what the shape or dimensions of the room actually are. You could be in a square room with walls just on the other side of the smoke, or you might be in a football stadium. Since your sight line is limited there's no way to tell.

Years after Kapteyn's work, it was found that the galaxy was much larger than he had thought. This was determined using infrared and radio observations; both wavelengths of light can penetrate the dust that obscures visible light. Careful mapping of gas clouds, stars, and dust has revealed the true nature of the Milky Way Galaxy, and it's a grand place indeed.

When you first visit a new city it's a good idea to take a tour. So let's take a stroll through the galaxy, starting near the Sun's current position, and moving along to see the sights. Remember, sometimes even well-lit neighborhoods can hide some dangerous characters, so don't be fooled by the beauty and apparent serenity of your surroundings.

THE SUBURBS

The most prominent feature of the Milky Way is its flattened disk of stars, gas, and dust, all of which orbit the center of the galaxy itself (similar to the way the planets orbit the Sun). The disk is 100,000 light-years across and roughly 1,000 light-years thick,* and is held together by its own gravity. It is composed of majestic, sweeping spiral arms, like a pinwheel. Spiral galaxies are fairly common in the Universe. Some are small and relatively obscure, and some are grand and huge, with well-defined arms. The Milky Way is one of the latter. In fact, very few spiral galaxies have been found to be bigger than the Milky Way.

The spiral arms are interesting. Because stars revolve around the center of the galaxy faster nearer the center (again, like the planets in the solar system), you might expect the spiral arms to eventually wind up like twine around a spindle. But they don't. They are not permanent, fixed features, like branches in a tree. Instead, astronomers think they are more like celestial traffic jams. On a city highway, a traffic jam isn't a fixed feature either; cars move in and out of the jam, but the jam persists. Similarly, as stars orbit the center of the galaxy they move in and out of the spiral arms, but because of a quirk in the way gravity behaves in a disk, the feature itself stays.

---

* Like the Earth's atmosphere, the galactic disk fades away slowly with height above (and below) the plane, so an actual thickness is hard to determine. It also depends on how you measure it; bright, massive stars tend to stick near the galactic plane, while lower-mass stars can reach great heights. So the thickness changes with what kind of star you are using to trace it.

Gas clouds orbit the center of the galaxy much as stars do. When a gas cloud enters a spiral arm, it hits that gravitational traffic jam and slows down. If another cloud enters the arm right behind it, the two will collide. This interstellar fender bender compresses them both, and when clouds compress, they form stars. The stars born in this way have a range of masses, from very low to very high. The highest-mass stars are bright, and light up the arms. However, these are the shortest-lived (see chapter 3 on supernovae), and don't live long enough to exit the spiral arms. Since the bright stars stay in the arms, the arms appear bright. Moreover, there are very few massive, bright stars compared to low-mass, dim stars, so the overall number of stars *in* the spiral arms is not much different from the number of stars *between* arms. It's just that between the arms there are few or no bright stars. This makes the arms more prominent than they otherwise would be.

So they're well-lit, pretty, and bustling with activity and traffic. And, like a busy section of the city, they have their dangers too.

For one thing, the fact that they are crowded is a major danger in itself. Not from collisions, though: the odds of any two stars colliding in a galaxy are incredibly low. In fact, assuming that stars are evenly distributed throughout the disk of the galaxy (a fair assumption), the odds of the Sun getting close enough to another star to even have their mutual gravitation affect each other is essentially zero! The average distance between stars in the disk of the Milky Way is huge: several trillion miles, while stars themselves are only roughly one million miles across. Imagine two flies in an empty box *five miles on a side*—what are the odds of those two flies even getting within a few yards of each other, let alone close enough to physically collide? That math works out to be the same for stars. On a human scale, the Milky Way is an incredibly empty place.

So stellar collisions will be extremely rare in the disk. As we've seen, though, you don't need to be all that close to a star for it to affect you. A supernova within a few light-years would fit anyone's definition of "bad." A gamma-ray burst (GRB) can be thousands of light-years away and still put the hurt on us as well.*

Stars are small compared to the vast distances between them in the galaxy. But some objects are bigger—a *lot* bigger. This ups the odds of an encounter significantly. Such a cosmic rear-end collision would darken our days on Earth . . . literally.

A CLOUDY FUTURE

When Kapteyn was counting stars, trying to figure out the shape of the galaxy, he had no idea that dust would mess up his statistics.

He *certainly* had no idea it could kill us all.

Stars make up about 90 percent of the normal mass of the Milky Way.† The remaining mass is made up of gas and dust strewn between them. That may not sound like much, but it adds up to a whopping 20 billion times the Sun's mass! That is a *lot* of litter, floating in the darkness of space.

Called the *interstellar medium,* or ISM for short, the majority of this material is in fact dark. It's cold—hundreds of degrees below zero

---

* Neutron stars can be dangerous too. Some have incredibly strong magnetic fields, quadrillions of times stronger than Earth's, which are generated inside the star and go out through the surface. A starquake—literally, like an earthquake on the star, but measuring a terrifying 30+ on the Richter scale—can shake the magnetic field violently, creating an ultra-mega-super-duper version of a solar flare. The energy released is enormous; in December 2004 such a flare from a magnetar *50,000 light-years away* hit the Earth and *actually had a measurable effect on our atmosphere.* Magnetars are difficult to detect and incredibly rare (only a handful exist in the Milky Way), but they may in fact be the most dangerous objects in the galaxy. They're the mob bosses of the Milky Way.

† Most of the total mass of the Universe is made up of *dark matter,* a name scientists have hung on a type of invisible matter about which very little is known. Its existence is inferred by its effect on the normal, visible matter in galaxies, and it makes up something like 85 percent of all matter in the Universe. More is being learned about it every day, and one of the biggest goals in modern science is to determine the nature of dark matter.

Fahrenheit—and mostly consists of hydrogen interspersed with heavier elements like helium, carbon, and oxygen. Some of it is dust, mixed in with the gas when it got blown out by giant stars and supernovae.

A lot of this material is smeared around the galaxy, like a layer of grime on a car's windshield. It's ethereally thin, with just a few atoms knocking around per cubic inch—the equivalent of a high-grade laboratory vacuum. But space is big, and even that small amount of matter adds up. If you go outside on a moonless summer's night in the northern hemisphere, you might see the band of the Milky Way high overhead. If you look carefully, along the constellation of Cygnus, the swan, you can see that the diffuse glow of the galaxy is split in two lengthwise by a dark swath called the Great Rift. That is the effect of dust in the Milky Way: it obscures the stars behind it, blocking their light from reaching us. Galactic smog, if you will.

Not all of the ISM is diffuse, though: some of it is clumped. After viewing the Great Rift in Cygnus, wait six months and go outside on a winter's night. Turn your gaze to Orion, the hunter. Below the famous three stars forming his belt you'll see three fainter, more tightly aligned stars making up his dangling dagger. The middle star of the knife is not a star at all; even through binoculars it takes on a fuzzy appearance. Through a moderate telescope you can see that it's actually a gas cloud, and in deep images with large telescopes its true nature is revealed: the great Orion Nebula is one of the largest complexes of gas and dust in the galaxy, with a total mass estimated to be thousands of times that of the Sun.

It's about 1,500 light-years away, yet visible to the unaided eye—it's *bright.** That's because it's a stellar nursery, the birthplace of thousands of stars. Many of these stars are massive, hot, and bright. In fact, a solid dozen stars inside the nebula will one day explode as supernovae (and then the nebula will get *very* bright). All the stars living out their lives inside the nebula light it up, making it brilliant and gaudy, the way the lights on Broadway illuminate the clouds above New York.

---

* At that distance, the Sun would be totally invisible to the naked eye; you'd need a telescope to see it at all.

The magnificent Orion Nebula is one of the most beautiful objects in the sky. It is the location of intense star birth, and is lit from within by a dozen high-mass (and short-lived) stars. Located 1,500 light-years away, it is easily visible to the naked eye.

NASA, ESA, M. ROBBERTO (STSCI/ESA), AND THE HUBBLE SPACE TELESCOPE ORION TREASURY PROJECT TEAM

Such star factories are scattered around the galaxy, but coincidentally the Orion Nebula, one of the largest, is pretty close on a galactic scale. If the Milky Way were a football field, the Orion Nebula would be only one yard away.

So just how close can nebulae get to us? Everything in the galaxy orbits the center, all at slightly different speeds and trajectories. It's possible that the Sun could pass very close to such a cloud, and in fact it gets even more likely when the Sun enters one of the spiral arms; as mentioned above, gas clouds pile up there. When the Sun moves into an arm, it's like driving along the highway and suddenly plunging into a fog bank.

What would happen to us if we slammed into such a cloud?

The effects of a collision are actually fairly complex, and depend on a lot of factors, such as how many stars are forming, how close the Sun gets to them, how long the Sun spends in the nebula, and the detailed structure of the nebula on small scales.

We can generalize a bit, though. For example, the core of the Orion Nebula is where most of the action is: several very massive newborn stars are busily spewing out light across the electromagnetic spectrum in vast quantities. One complex of massive stars at the very heart of the nebula cranks out as much energy in *just* X-rays as the *total* energy the Sun emits! Even so, it would take a very close passage to these stars to affect us on Earth; even from a light-year or two away the X-rays from them would affect us far less than an average solar flare.

The ultraviolet emission isn't too big a deal either. The brightest young star in the heart of the Orion Nebula is named Θ1c Orionis,* and it has a mass 40 times the Sun's and a surface temperature 7 times hotter. Ultraviolet light floods out of such a star; Θ1c's ultraviolet output is *millions* of times that of the Sun. However, from a light-year away that emission is diluted greatly, and we'd receive only a fraction more UV than we do from the Sun.

In addition, Θ1c blows out a stellar wind, and it's beefy: it spews out 100,000 times as much matter as the Sun does in its solar wind, and at twice the velocity. However, again, from a light-year away the wind would be attenuated enough that the Sun's magnetic field would protect us from the onslaught.

The most dramatic effect would be the visible one: from a light-year away, the brightest stars in the Orion Nebula would be incredible to see—Θ1c blasts out energy at a rate over 200,000 times that of the Sun! From a light-year away it would shine almost as bright as the full Moon.

---

* Pronounced "thay-ta one cee ore-ee-ON-us," if you want to impress your friends.

Dark clouds of dust haunt the Milky Way. The density of particles in them is very low compared with our atmosphere, but the clouds are so huge that they are opaque. They absorb the light from stars behind them, leaving what looks like a great hole in the sky. If you look carefully at this one, called Barnard 68, you can see the stars getting dimmer as you gaze from the outside of the cloud toward the center.

EUROPEAN SOUTHERN OBSERVATORY

Other stars in the nebula would also be incredibly bright, and scattered through the sky; a truly dark night would be virtually unknown. This might affect some nocturnal species (see chapter 3) but overall it wouldn't be too big a deal.

That's not to say that a nebula is a cozy place to be. Perhaps the most dangerous aspect of passing close to the center of the Orion Nebula is that it would take a long time. Stars like Θ1c have the unfortunate tendency to explode, with all the dangers involved (again, see chapter 3). Supernovae are dangerous within about 25 light-years—closer than that and the explosion does serious damage to the Earth's ozone layer, causing a potential mass extinction. A close pass through the heart of the nebula means the Sun will be in the danger zone for close to 100,000 years.* Massive stars live short lives of only a few million years before they explode, so there is a significant chance that plowing through a nebula like Orion will bring us dangerously close to an exploding star. Just one more fun thing to think about.

And there are two more dangers in this close encounter, both of which are invisible. Or not invisible so much as *dark*.

So far, I've only talked about beautiful nebulae illuminated by their newborn stars. But not all nebulae are like that; some have not yet formed stars. These are dark, cold clouds that go by various names, such as *molecular clouds, Bok globules,* or simply *dark nebulae.*

Some of them are fairly dense as cosmic objects go, with as many as 100 million particles of dust per cubic inch. To be sure this is still not terribly dense; Earth's atmosphere at sea level is a hundred billion times denser! But these clouds can be very large, light-years across, and that adds up. Like a thick fog, they can completely absorb any starlight that falls on them. Many of them look almost like holes in space, so completely do they block light.

Interestingly, the exact effect on the Earth is difficult to predict were the solar system to plunge into such a cloud.† Certainly, the amount of sunlight reaching the Earth could drop significantly; even a few percent diminution of sunlight could start an ice age. There are defi-

---

* Assuming the Sun's velocity through the nebula is the same as its orbital velocity around the galaxy of 140 miles per second.

† We wouldn't actually feel it, I'll note, since even the thickest nebula is incredibly rarefied. The Earth wouldn't slow in its orbit or anything like that. You'd hardly notice, except for the effects outlined above.

nitely dark nebulae in the galaxy dense enough to block that much sunlight.*

And what of the dust that physically mixes into Earth's atmosphere when we plunge into a dense nebula? A group of scientists investigated what would happen to the Earth if this occurred, and they found that dust can accumulate in the Earth's atmosphere, enough to darken the skies and significantly lower the Earth's temperature. It could even cause a runaway ice age. They also determined that moderate ice ages can be triggered by less dense clouds, which we encounter somewhat more often. They estimate that we encounter such a cloud about every 100 million to 1 billion years or so, which means it's a dead-on certainty that this has occurred several times in the Earth's history. It's probably happened a few times since complex life evolved on Earth too, though no specific ice age on Earth has been positively identified as having been triggered by a collision with a dark cloud.

However, there is another danger from getting too close to a nebula, and this time the details of the cloud aren't so important. All that matters is, well, its *matter.*

Some interstellar clouds are incredibly massive, hundreds of thousands or even millions of times the mass of the Sun. A nearby passage means we will be affected by the gravity of all that mass. The direct effects on the Earth are minimal, actually, since we are so close to the Sun that its gravity will dominate.

But not all objects in the solar system are safely nestled in the inner solar system. Surrounding the Sun, well beyond the orbit of Pluto, is the so-called *Oort cloud* (named after the Dutch astronomer who postulated its existence), a vast collection of giant chunks of ice and rock, some of which can be hundreds of miles across. Some of these icebergs

* There might be a mitigating factor: the Sun will heat up the dust surrounding it, which will in turn warm up the Earth. The exact details of this, though, are difficult to calculate, and depend on lots of niggling factors, such as the density of the cloud, its composition, and all that. Would the warm dust offset the darkening Sun enough to stop the glaciers from advancing? We simply don't know.

have orbits that bring them into the inner solar system every few dozen millennia, and when one of them comes, we see it as a beautiful comet.

Oort cloud objects typically stay well away from the Sun, hundreds of billions of miles out. It takes some sort of perturbing influence, some kind of shove, to change their orbits enough to drop them into the inner solar system. Such an effect may come from a passing star a few light-years away, for example; at the distance from the Sun of a typical Oort cloud object it takes just the thinnest whisper of a nudge to send them down.

If the Sun strays too close to a giant nebula, that whisper can turn into a shout. Some estimates of the Oort cloud put its population of orbiting icebergs in the *trillions.* Go back to chapter 1 and read about the damage a comet or asteroid impact can do. Now multiply those effects by ten, or a *hundred,* as comets rain down from the heavens after a close passage with a massive nebula.

Yikes. It's hard to imagine the devastation wreaked by such an event. The Earth's biosphere might just start recovering a few centuries after an impact when another comet would slam into us. How many mass extinctions in the dim history of our planet were due to the Sun skirting too close to a giant gas cloud?

It's ironic—the Sun was almost certainly born in such a gas cloud 4.6 billion years ago. It may have once been surrounded by massive stars littering the sky, their stellar winds creating vast shock waves across the gas, compressing it into sheets and filaments that glowed like neon signs crisscrossing the sky.

Heading into such a gas cloud might almost be worth it. What a view!

But then again, a nice dark sky with all those nebulae at a safe distance of a few thousand light-years away sounds pretty good too.

PLANE FLIGHT

As mentioned above, the stars in the disk of the Milky Way orbit the galaxy's center similarly to the way the planets orbit the Sun. However, there are some important differences. On the scale of the solar system (many billions of miles across), the Sun is small (less than one million miles across). As far as the planets can tell, all the gravity in the solar system is concentrated in one spot.* Because of this centrally located source of gravity, the orbit of a planet can only have a certain kind of shape, called a *conic section.* This includes circles, ellipses, parabolas, and hyperbolas. All of these shapes are planar; that is, they are flat. If you smack a planet hard enough the orbit will change shape, or it might change the tilt of the orbit, but the orbit itself will still be a conic section, will still be flat.

But the situation is different for stars orbiting the center of the Milky Way, because the mass is spread out, distributed around the disk. A star orbiting in that disk feels gravity from masses all around it, and not just from a single point in the galactic center. Orbits of stars can therefore have all sorts of weird shapes. Let's say you have a star that orbits the galaxy in a perfect circle that is exactly in the midplane of the disk. If you were to give the star a little bit of vertical velocity—perpendicular to the disk—the star would bob up and down relative to the disk, like a cork floating on water (while still circling the center).

It's a little like throwing a rock up in the air; gravity slows it and it falls back down. The vertical velocity of the star propels it above the plane of the disk, but the disk's gravity pulls it back down. The disk, though, isn't solid; it's made up of stars that are separated by large distances. There is nothing to stop our star, so it passes right through the plane, and heads down, below it. Again, the gravity slows it to a stop, and the star reverses course. The cycle will repeat forever if the circumstances are right. When you couple this with the star's circular orbit, you get a shape like a sine curve wrapped into a circle.

There are many ways a star could get started on an excursion like this. It could pass by another star, and the gravitational interaction

---

* More or less, that is. The planets themselves do have gravity, and they do affect one another, but only very subtly and only on very long time scales. We'll be returning to this idea in a moment.

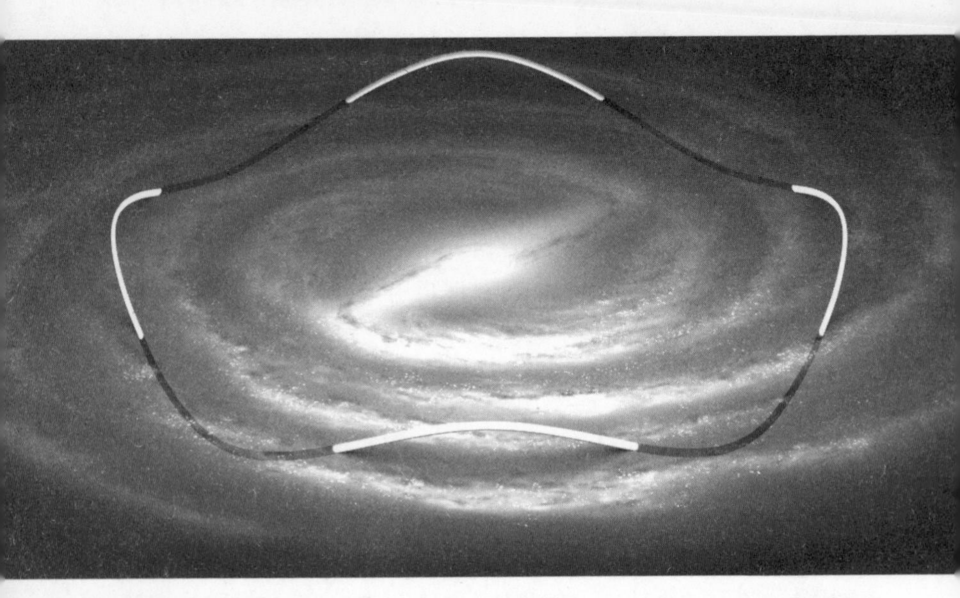

Unlike planets orbiting the Sun, the Sun itself bobs up and down as it circles the center of the Milky Way. It pokes up above the disk about every 64 million years, making roughly four cycles every time it orbits the galaxy once. The vertical scale has been exaggerated here; the amplitude of the Sun's motion is really only a few hundred light-years up and down.

NASA/JPL-CALTECH/R. HURT (SSC) AND CHRIS SETTER, B.I.L.

could kick the star upward or downward—but as we saw before, stellar encounters are extremely rare, so this is unlikely. On the other hand, stars form in clusters (see below), where they are much closer together and gravitational interactions are more common. A massive star in the cluster passing close to a less massive one could easily toss the smaller star right out of the cluster, and also impart a bobbing motion.

Another way is for the star to pass near a giant cloud of gas and dust. We saw above that a direct collision with a nebula has some deleterious effects, but another is that the mass of the cloud can warp the orbit of the star, giving it a vertical velocity and forcing that bobbing oscillation.

It turns out that a star very near and dear to us exhibits just this sort of motion: the Sun! Careful measurements of the Sun's velocity relative to the stars around it show that the Sun is in fact oscillating above and below the galactic plane. The excursion isn't huge: maybe 200 light-years or so at maximum compared with the disk's diameter of 100,000 light-years. The disk is also about 1,000 light-years thick, so the Sun still stays within the bulk of the material of the disk as well.

The period of the Sun's oscillation—from maximum height above the disk, diving down through it to the maximum depth below the plane, then back up to maximum height—is about 64 million years.

Well, that sounds cool: we get a free ride to a (slightly) better view of the galaxy over a few million years, and no harm done, right?

Right?

Maybe not. But to see why, instead of looking up, we have to look *down,* into the layers of sediment on Earth.

For many years, it's been suspected that the fossil record of life on Earth has shown a periodicity in mass extinction events, as if life on Earth is following some sort of schedule for huge die-offs followed by a rediversification of species. Not all of these events follow such a schedule, and for many of them a smoking gun has been found; the most famous is the end of the dinosaurs, and we have pretty good evidence that an asteroid impact was behind it. But for others (with the exception of perhaps the Ordovician extinction event; see chapter 4) the causes aren't so clear.

A periodicity to mass extinctions implies some sort of cyclical cause, of course. While it's impossible to rule out things like episodic super-volcano eruptions or some other internal cause, cycles on really long time scales imply extraterrestrial forces.

Until recently, this cyclical large-scale grim reaping has only been suspected; the fossil record wasn't all that clear. But new research has strengthened the supposition considerably. By mathematically analyzing the fossil record, researchers have discovered a very strong signal of periodicity in the mass extinction history. They examined diversity of species—literally, how many species of life existed at different points in the fossil record—and have found that the number of different species appears to rise up and down with a distinct period.

That period, they determined, is about 62 million years.

Uh-oh.

Is it merely a coincidence that cycles of extinctions match up closely with the period of the Sun's oscillation into and out of the Milky Way's disk? There are ways to check, statistical methods to try to match up two different cycles and see if they might be correlated. Another group of researchers, Mikhail Medvedev and Adrian Melott at the University of Kansas, carefully performed this analysis, and their answer is "maybe."

Well, that's not terribly reassuring. But this is a new field of research, and we're just getting started looking into it. The data are sparse, and the results so new it's hard to say how firmly based the conclusions are.

But they are certainly provocative.*

In this case, the culprit may be our old friend the cosmic ray. As you might remember from previous chapters, these little guys are subatomic particles accelerated to enormous velocities in outer space. When they impact the Earth's atmosphere, there are a number of effects. For one, when a cosmic ray smacks into a molecule of air at nearly the speed of light, it shatters into a shower of smaller subatomic particles called muons. These scream down from the sky, and if they hit a DNA molecule in a cell they can alter or destroy it. This actually happens all the time, but in general living tissue can repair or reject the damage. But if enough muons rain down, there could be slow but long-term effects on life—a mass extinction, for example. As noted earlier, muons can penetrate water to depths of more than a mile and also go right into rock down to depths of half a mile. This would therefore affect nearly all life on Earth.

Cosmic rays have other effects as well. They can destroy ozone molecules in the upper atmosphere, exposing life below to dangerous levels

---

* Provocative in the literal sense as well, since these findings have provoked a flurry of papers both supporting and attacking their conclusions. I want to stress again that this periodicity in mass extinctions has not been verified, and may in fact not be real. Time will tell as more work is done.

of ultraviolet light from the Sun. They can also create nitrogen dioxide, which can form acid rain. Over years, this can destroy plant life, and this effect would work its way up the food chain.

Finally—and perhaps less well established—cosmic rays can seed cloud formation, so an increase in cosmic-ray influx may increase the amount of cloud coverage on Earth, forcing climate change as more sunlight is reflected into space. While it may not necessarily incite a full-blown ice age, even a temperature drop of a few degrees can be devastating to the biosphere.

But where do these cosmic rays come from? And how is this tied into the Sun's bobbing motion as it circles the galaxy? If such a connection does in fact exist, Medvedev and Melott may have found it.

Most cosmic rays come from supernova explosions and pulsar winds; the material moving outward from those sources can slam into slower material and generate fierce shock waves that accelerate subatomic particles like protons and electrons to within a razor's edge of the speed of light. Because they originate from events happening inside the Milky Way, they are called *galactic cosmic rays.*

But there are cosmic rays that come from outside the galaxy as well. The Milky Way is part of a small cluster of galaxies called the Local Group, which consists of our galaxy, the Andromeda galaxy (a massive spiral similar in size to ours), and a handful of smaller galaxies. The Local Group is on the outskirts of the much larger and far more massive Virgo Cluster, which contains *thousands* of galaxies—we're like the suburbs of a vast metropolis. The Virgo Cluster's gravity is not to be trifled with: we (and the other Local Group galaxies) are locked in its grip, and being pulled toward the cluster at the astonishing speed of 160 miles per second.

And we're not moving in a vacuum. Remember the *intergalactic medium*? The Milky Way slams into this rarefied stuff at high speed, creating a shock wave almost beyond imagining: it's hundreds of thousands of light-years across, and generates *huge* amounts of energy. The energies are so vast that they create cosmic rays, but in this case they come from outside the galaxy, so they are *intergalactic cosmic rays.* The cosmic rays scream away from the shock front, and many of them are aimed our way, back into the galaxy.

The galaxy, like the Sun, has a magnetic field. Also like the Sun's, the galactic magnetic field is a mess of twisted, coiled loops. They are strongest right in the middle, the midplane of the disk, where the magnetic field does an excellent job of deflecting incoming galactic cosmic rays. However, their strength dims rapidly with height above or below the plane. If a star stays near the plane, it is protected from these high-energy particles. If it strays too far, the star gets exposed to them.

And this is where the oscillating Sun comes in. Bobbing up and down, above and below the plane as it orbits the center of the Milky Way, the Sun finds itself high above the plane and its protective magnetic fields every 64 million years. This is toward the direction of the cosmic shock wave, where the Sun is relatively unprotected from the incoming cosmic rays. It's like facing upwind while a tornado flings gravel at you. Medvedev and Melott found that the number of intergalactic cosmic rays that can reach the Sun during these periods can increase by *a factor of five* over quieter periods when the Sun is far below the galactic plane (which also has the shielding effect of putting the bulk of the galaxy between us and the incoming cosmic rays).

The number of intergalactic cosmic rays that can reach the Sun thus goes up and down significantly over time with that 64-million-year cycle. The scientists then used the Sun's predicted motion to make a model of the number of cosmic rays reaching us here on Earth, and plotted it against the graph of the fossil diversity going back in time. They found that *the maxima from the first plot overlaid the minima from the second plot* every *time!*

In other words, whenever the Sun was high above the plane, and the number of incoming cosmic rays was at its peak, the number of species of life on Earth decreased. *Every single time,* back over the past nine cycles, over half a billion years.

Let's be clear: this is not direct evidence that the Sun's motion causes mass extinctions. But it's very compelling. When the researchers accounted for asteroid impacts and other non-cosmic-ray events that cause mass extinctions, the correlation between the Sun's motion and those massive die-offs got even better. Incidentally, the research doesn't indicate precisely what it is about cosmic rays that delivers the blow. There is some evidence that ice ages are also correlated with these periods, so perhaps cloud cover and climate change are behind it all. There's interesting research connecting cosmic rays with the triggering of lightning on Earth too. It's not clear which of the ways outlined above (muons, ozone depletion, smog generation, or cloud seeding) does the dirty deed, or if it's a mix and match of any or all of them, or maybe something we haven't even guessed at yet. But there is mounting evidence that cosmic rays do have an effect on life on Earth.

This raises the obvious question: where are we now in the cycle? The Sun is currently on its way up, above the disk. We are only 25 or so light-years above midplane, well protected by the galactic magnetic fields, so we have a ways yet to go before we're in the danger zone. Our descendants 20 or 30 million years from now, however, may have cause for concern; they'll be watching their neighborhood deteriorate. If they can avoid the ever-heating Sun, supernovae, and the odd gamma-ray burst or two, they may still have intergalactic cosmic rays to deal with. To avoid them, they'll have to find some Sunlike star with habitable planets in the galactic midplane (or below it) and move there. There are probably lots of potential colony sites . . . if the stars are right.

MONSTER IN THE MIDDLE

Our great-great-great (etc.) grandchildren may have another problem besides intergalactic cosmic rays. This one is slightly closer to home—just downtown, as a matter of fact. To understand it, though, we'll need to take a small step back in time, and a giant leap back in space.

In 1963, astronomers had an enigma on their hands. Radio astronomers had discovered an object that was pretty bright in the radio part of the spectrum, which is always nice. The problem was, the technology of the time wasn't up to nailing down the object's position very well—

similar to the problem gamma-ray-burst astronomers would face a few years later.

A cosmic coincidence saved the day: the object—called 3C273—is in a location in the sky that happens to overlap the apparent position of the Moon as it orbits the Earth. This means that every now and again, the Moon appears to pass right over 3C273, blocking it from our view. By timing precisely when the radio waves from the object are blocked by the Moon's sharp edge, and knowing the exact position of the Moon, they were able to determine the object's location with high accuracy . . . and when they trained their optical telescopes on that position, all they saw was a faint blue star. This was quite shocking—how could such a feeble emitter of visible light be so luminous in radio?

Things got even more perplexing when the distance to the object was found to be a staggering *one billion* light-years. Far from being an innocuous faint blue star, 3C273 must be *the most luminous known object in the Universe.*

Soon more objects like this were found, and they were dubbed *quasars,* for "quasi-stellar objects." Other, similar objects were found as well, and sported names like *blazars* and *Seyferts.* They all emit light across the spectrum, and some are true monsters, emitting many trillions of times the Sun's energy, hundreds of times the total energy output of our entire galaxy!

Over time, it became clear that these objects were galaxies similar to ours, except that energy blazed forth from their cores, making them incredibly luminous. What could do that? Whatever the source of that energy was, it had to be small,* and it had to produce radio, optical, and X-ray light on vast scales.

---

* The size of the actual energy source was known to be small because of some complicated physics involving how rapidly the source changed brightness—the bigger it is, the slower it can vary its output. Rapid fluctuations in the energy emission from 3C273 and other quasars made it clear that the source of their prodigious energy must be on the same scale as our solar system—tiny when compared to an entire galaxy.

M87, a giant elliptical galaxy in the Virgo Cluster, is the nearest active galaxy. The supermassive black hole lurking in its core emits a giant jet of energy and matter moving at nearly the speed of light.

NASA AND HUBBLE HERITAGE TEAM (STSCI/AURA)

Only one object astronomers knew of fit all these characteristics: a black hole.

But even stellar mass black holes couldn't put out that kind of power. Astronomers came to grips with the fact that there must be a different kind of black hole, a far scarier kind: a *supermassive* black hole (SMBH).

In fact, over time it was found that every large galaxy in the Universe has an SMBH at its core. Even our Milky Way does—it's called Sagittarius A* (pronounced "Sagittarius A star"), or Sgr A* for short—tipping the cosmic scales at 4 million times the Sun's mass.

And it's considered a lightweight. The central black hole in the giant elliptical galaxy M87, which at 60 million light-years distant is much closer than 3C273 (though that's still a long walk), has one of the most massive SMBHs ever seen, weighing in at one *billion* solar masses.* These superluminous objects—now collectively called *active galaxies*—are so bright because the black holes in their cores are actively feeding. Material, gas, dust, and even stars are falling into the gaping maws of these monsters. As the matter falls in (similar to when a black hole forms in a gamma-ray burst) it forms a flattened accretion disk. Friction and magnetic force heat the disk to millions of degrees, and matter that hot gets very, very bright (see chapter 5). It will emit numbing amounts of light, dwarfing the combined light from the rest of the galaxy. It will also emit X-rays and even gamma rays, the highest-energy form of light.

As we saw in chapter 4, a black hole with a disk can also form jets of matter and energy, and supermassive ones can do this as well. Not all active galaxies' SMBHs have jets, but many do. It's like a GRB on a galactic scale, but instead of a few-seconds-long flare of energy, the jets are stable, constant sources of power, lasting for millions of years or longer. Active galaxies are the largest reservoirs of energy in the Universe.

The environment inside one of these active galaxies must be interesting, if by "interesting" you mean "terrifyingly scary beyond belief." Even without a jet, the core of these galaxies would be booming out energy across the electromagnetic spectrum. Any star near the core would be bombarded by radio waves, optical light, X-rays, maybe even gamma rays. It's hard to imagine life being able to arise on a planet orbiting a star near the center of an active galaxy.

Even other galaxies may not be safe from such an unfriendly neighbor: 3C321 is a pair of galaxies, one of which is active. The active one is shooting out a jet directly at its partner 20,000 light-years away. The beam is creating all kinds of havoc in the victim galaxy, including ramming the clouds of gas there, irradiating the stars, and generally ruining

---

* Some very distant quasars have SMBHs estimated to have as much as 10 billion solar masses, but these have yet to be confirmed.

what was probably a pretty nice neighborhood before all the mayhem started.

Which brings us to an interesting juncture. Can the Milky Way become an active galaxy? Can the galaxy *itself* become a danger to us?

In fact, yes it can. And it probably has been one in the past.

At the moment, the Milky Way's black hole is napping—it takes incredibly sensitive gamma- and X-ray detectors to see any emission from it at all. For an SMBH to be active, a lot of material must be falling into it. Evidently ours is either not eating or not eating very much. We do see *some* energy coming out, but it's very diffuse and very faint. Astronomers aren't sure what's causing this emission, and that very uncertainty of the source indicates that the Milky Way is not a booming active galaxy (or else the source would be obvious). So we appear to be safe.

But appearances can be deceiving. Studies have shown that there is quite a reservoir of gas near the black hole. Stars in the vicinity emit particle winds like the solar wind, and this matter can accumulate near the black hole, feeding it. These same studies show that the stream of particles can become clumpy, and when a big clump falls into the black hole, it can suddenly flare, becoming active for short periods. It emits vast energies for a few years before settling down again. These flares are most likely not very dangerous to us; the last one may have been as recent as 350 years ago—its effects are imprinted in the gas surrounding the galactic center, which can be more easily seen. X-ray observations of these clouds indicate that the last flare emitted energy at a rate 100,000 times higher than when the black hole is quiet. This sounds frightening, but remember, this happened recently as astronomical effects go, and humanity didn't even notice.

Remember too that we're located 25,000 light-years from the galaxy's center, which is a lot of real estate between us and it. So it appears we're not in any danger from such flares.

However, there are other reservoirs of gas near Sgr A*. Vast dark clouds of gas with more than a million times the mass of the Sun lurk nearby. They are currently stably orbiting the galactic center . . . currently.

When galaxies collide, beauty (and terror) can result. This galaxy, called the Tadpole because of its shape, had a recent encounter with another galaxy. The gravitational dance of the collision drew out a long streamer of gas from the Tadpole. In many such collisions, gas can be dumped into the centers of the galaxies, causing them to become active.

NASA, H. FORD (JHU), G. ILLINGWORTH (UCSC/LO), M. CLAMPIN (STSCI), G. HARTIG (STSCI), THE ACS SCIENCE TEAM, AND ESA

If you look at images of active galaxies, you might notice a trend: a lot of them are, well, *funny*-looking. They are distorted from the usual spiral or elliptical shape. Astronomers think this may be due to recent encounters with other galaxies, traffic accidents on a truly galactic scale. When two galaxies collide, their gravitational interaction can cause gas and dust to stream into their centers, where any supermassive black hole will eagerly gobble it up. This, in turn, will switch on the black hole, turning the recently quiet galaxy into an active one.

The Milky Way is not immune to such things. It has eaten many smaller galaxies in the past; in fact, it's likely that most or even all large galaxies have grown through cannibalizing their neighbors. These types of encounters would have been more common in the past, when the Universe was smaller and galaxies were closer together. In fact, objects like quasars are all very far away, which means we see them when they were younger, in the past.* It was a galaxy-eat-galaxy Universe back then, and it's possible—even likely—that *all* major galaxies, including our own, were once active in their youth.

Encounters in recent times are more rare, but not unknown. The Milky Way is currently ingesting at least two different small galaxies, but these events are far too small to activate our SMBH. There are currently no nearby galaxies big enough and close enough (at least for now; see below) to do the deed, so most likely we're safe from our own local active galaxy.

Of course, it's *possible* that two clouds on different orbits around the black hole could collide, canceling each other's momentum, sending them down into the monster's maw. If that happened, the black hole could switch on and stay active for millennia, flooding the galaxy with vast levels of X-rays and streams of subatomic particles like a firehose on a cosmic scale.

The good news there is that this emission would be beamed, like a gamma-ray burst. Most likely, the beams would head up and down, out of the Milky Way's plane and away from us. If that's the case, we're safe enough.

Of course, some galaxies have black holes in which the axis is tilted with respect to the plane, so it's *possible* their beams could actually plow

---

* Because light travels at a finite speed, we see a distant object as it appeared in the past. It takes light 8.3 minutes to get to us from the Sun, so we see it as it was 8.3 minutes ago. We see a galaxy 10 billion light-years away as it was when the Universe was very young, only a few billion years old. In effect, telescopes are time machines. In reality—and as usual when dealing with relativity, time, and space—the situation is more complicated than this, but it's not terrible to think of distance (in light-years) as equal to time (years in the past).

through the stars in the plane. But those are rare, and even if the Milky Way's SMBH were one of them, the odds of a beam's hitting us are probably only 1 in 30 or so.

I'd prefer longer odds myself, but then the series of events needed for us to be looking down a gamma-ray beam from the supermassive black hole in the Milky Way's heart are already pretty precarious. I think we're fairly safe.

And before you get too biased against supermassive black holes and their destructive powers, consider this: they may be necessary for life to arise.

Since every galaxy has a big black hole in its center, there is some reason to think that black holes play a role in galaxy formation. In fact, some characteristics of galaxies—like the way stars orbit the galaxy's center—seem to scale with the central black hole's mass. You might think that's natural given how big the central black hole is, but remember: even a billion-solar-mass black hole is only a tiny fraction of the mass of a galaxy! The Milky Way is at least 200 billion solar masses, so our own supermassive black hole harbors only 0.002 percent of the total mass.

Theories abound, but it looks like the supermassive black hole in each galaxy formed at the same time the galaxy did. As stars formed and the matter forming the galaxy streamed into the center, the black hole accreted mass, becoming active, and blew out huge winds of particles and energy. These winds must have profoundly affected the galaxy around it, possibly even curtailing the size of the galaxy itself as it was forming. They would have influenced star formation, and the chemical content of those stars as well.

Sure, black holes can kill us, and in a variety of interesting and gruesome ways. But, all in all, we may owe our very existence to them.

Remember: when you stare into the abyss, sometimes it stares back at you.

ANDROMEDA STRAIN

There's one more stop on our galactic tour, and technically it's not really a danger from our own galaxy. But it involves the Milky Way, and honestly, it's just too cool not to spend a moment on.

As mentioned earlier, our galaxy is not alone. Like a city surrounded by towns, several smaller galaxies hang out in our Local Group. But there's also another big galaxy in the Local Group: the Andromeda galaxy. It's a bit more massive than the Milky Way, so it's the Minneapolis to our St. Paul (or the Baltimore to our Washington, D.C., or the Dallas to our Fort Worth, or whatever other cartographical analogy you like). Between the two of us, we totally dominate the Local Group.*

Estimates vary, but the best guess is that Andromeda is about 2.5 million light-years from our own galaxy. Because the two galaxies are each about 100,000 light-years across, this makes them unique in terms of scale: the distance between them is not that much bigger than their size. Stars are incredibly far apart compared to their sizes, as are planets. But galaxies are big, and can be close together . . . and that means they can interact.

Astronomers have measured the relative velocities of the two galaxies, and it looks as if the pair are bound together by their mutual gravity. In fact, there's an even stronger sign that the two galaxies are doing a do-si-do.

As far as we can tell, almost all big galaxies in the Universe appear to be rushing away from us. The details of this aren't important here—they'll be in the next chapter in spades—but this means that over time, every big galaxy in the Universe will move away from us . . . except for one. You guessed it: Andromeda. That nearest big spiral is unique in the heavens because it is actually headed *toward* us.†

---

* There is one other spiral in the group, called M33 or the Pinwheel galaxy. Although it's a spiral like the Milky Way and Andromeda, it has only a fraction of the mass, so it's not a big player like us.

† Actually, many of the other galaxies in the Local Group are bound to us as well, but again, they are much smaller.

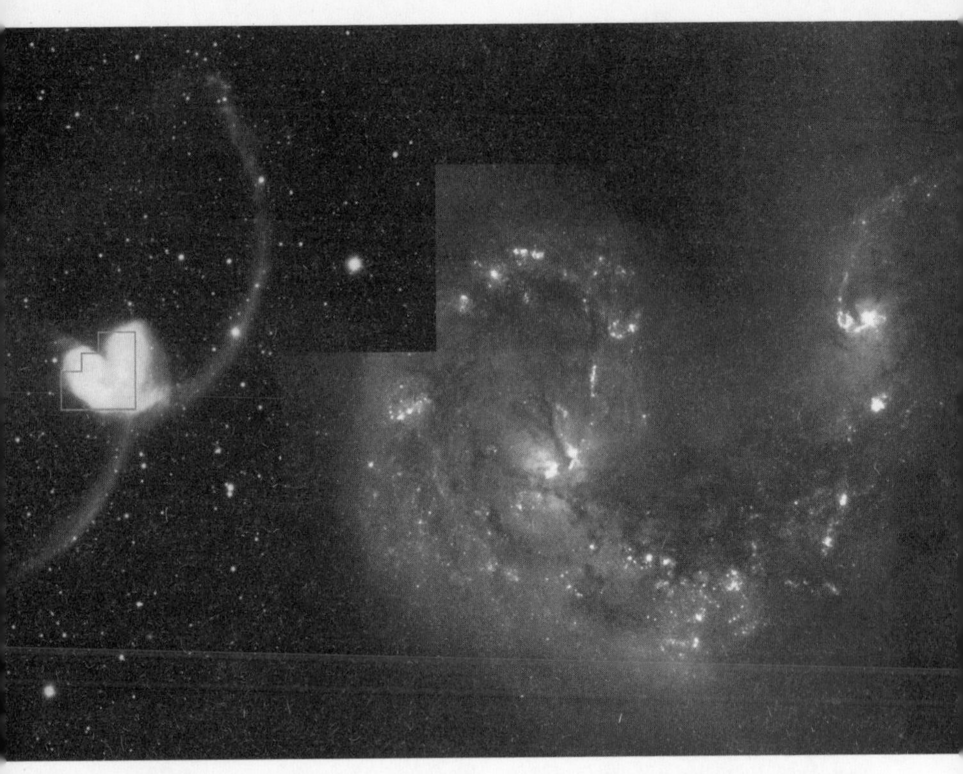

The Antenna galaxies (so called because of the long, curved antennae of gas and stars protruding from them) collided millions of years ago, and are in the process of merging. Their gas clouds are colliding on epic scales, causing massive amounts of star formation. Any spectators in those galaxies would have a fantastic view . . . for a while.

BRAD WHITMORE (STSCI) AND NASA

In point of fact, it's *screaming* toward us: its velocity toward the Milky Way is about 120 miles per second, which is pretty fast (keeping up with our city theme, during the time it takes you to read this sentence, the Andromeda galaxy would have covered the distance from New York City to Boston). The problem is, we don't know exactly what its *transverse velocity* is, its motion *sideways* relative to us. Think of it this way: if you're standing in the street and a car is headed at you, that's bad. But if it's also skidding to the side quickly enough, it'll miss you.

We don't know for sure how much Andromeda is moving to the

side. At its current distance from us, even a transverse velocity of hundreds of miles per second translates to a very tiny shift as seen by a telescope. However, it's safe enough to assume that the transverse velocity is roughly the same as its velocity directly toward us, and some theoretical models back that up. That's not enough for it to totally miss us.

So, given enough time, Andromeda and the Milky Way are due for a train wreck. What will happen?

Two astronomers decided to find out. T. J. Cox and Abraham Loeb at the Harvard-Smithsonian Center for Astrophysics modeled the interaction between the two giants over several billion years. What they found out doesn't bode all that well for us.

The two galaxies accelerate toward one another as they close in. Faster and faster they approach, until they finally physically collide about two billion years from now. The collision is almost ethereal—stars are so far apart that in essence the two galaxies will pass right through one another. The odds of any two stars getting close enough to physically collide are practically zero.

In most galaxy collisions we observe today, the victims are suffering a burst of star formation.* This is because gas clouds, unlike stars, are very large, so in a typical galaxy collision the chance of a *cloud* collision—of *many* collisions—is a virtual certainty. When the clouds collide, they collapse and form stars. Many of these stars are massive and hot, so they light up the gas around them. Galaxy collisions in the Universe today advertise their presence by lighting up like neon signs.

However, according to the model created by Cox and Loeb, by the time the Milky Way and Andromeda merge a few billion years from now, much of the gas currently existing in the two will have already been used up to make stars. Unlike other galaxy collisions, our own

---

* To be specific, this should say "collisions between *large* galaxies." Big galaxies eat small ones all the time; the Milky Way is cannibalizing two dwarf galaxies right now. Both have been torn apart by the galaxy's gravity, and their stars are slowly becoming integrated with the original Milky Way population. This has happened many times in the past as well.

won't be accompanied by a starburst. This makes the collision safer for us; no starburst means no giant clusters of massive stars irradiating their environment, and no wave of supernova explosions destroying everything around them.

That doesn't mean there's no drama, however. During the collision, the shapes of the galaxies get distorted. Currently the Milky Way and Andromeda are both "grand design" spirals, with majestic spiral arms. But imagine you are a star on the galaxy's edge, on the side facing Andromeda. As the other galaxy nears, you start to feel a gravitational tug from it, and eventually that pull is equal to the force you feel from your home galaxy. A star on the far side of the Milky Way, however, feels a greatly reduced pull since it is so much farther away from Andromeda. This has the effect of stretching out the galaxies, pulling them apart like taffy, forming long tentacles called *tidal streams.*

Over millions of years the two galaxies pass each other, whipping around in a curving path (depending on the amount of transverse velocity). The two long tails of stars, gas, and dust pulled out from the galaxies curve along with them, forming glowing tentacles hundreds of thousands of light-years long. From some distant galaxy, the two would look like some weird pair of marine creatures fighting to the death (or perhaps mating).

While the two galaxies pass through each other, they don't have enough velocity to escape each other's grasp. After about another billion years they fall back toward one another, repeating the sequence, and then again in less than another billion years. Finally, about five billion years from now, the two galaxies will have merged. Their cores will coalesce, and the matter ejected into the long tails will settle into a stable orbit. Instead of two spirals, the resulting merger will yield a single giant galaxy that is elliptical in shape—Cox and Loeb have dubbed it *Milkomeda* (I suppose *Andromeway* sounded too much like the name of some sort of pharmaceutical). In fact, many of the giant elliptical galaxies seen in the sky may be the result of such massive mergers; they are the junk heaps of cosmic collisions.

But what of the Sun? What happens to us during all this?

Interestingly, this whole event transpires during the lifetime of the Sun. While the Sun may be a red giant by the time it all ends (see chapter 7), it'll still be around. Maybe.

Cox and Loeb's model can make some predictions about the Sun's fate. They find that after the first passage of the two galaxies, the Sun has a large chance of staying within the Milky Way's disk. However, there is a small chance (about 12 percent) that it will be ejected into one of the long tidal tails. There is no danger from this, and in fact (as we'll see in a moment) this may be the safest place for us to be. And the view! From that vantage point, we'll be looking down on the collision with very little dust to obscure the scene. We'll have box seats to one of the most colossal events in the Universe.

The chance of the Sun's getting tossed out of the Milky Way becomes greater with each passage of the two galaxies. By the time the cores merge, the odds of the Sun's being farther than 100,000 light-years from the center of the merger remnant are about 50 percent (and we're better than 3 to 2 to be at least 65,000 light-years from the center). We're currently about 25,000 light-years from the Milky Way's center, so that's a significant change.

In fact, during the merger there is a small chance (less than 3 percent) that we'll swap sides, becoming bound to the Andromeda galaxy! While these are long odds, it's an amazing idea. Stars tend not to be fickle in collisions, and stick with the ones who brought them, but a few will change allegiance given the chance.

There is also another possibility: there is a small chance—less than 1 percent, but it's there—that the Sun will actually drop toward the center of the system. If this were to happen, then the Sun could actually get within a few thousand light-years of the merged cores of the two galaxies, and this would be very, very bad.

Remember, all large galaxies have supermassive black holes in their centers. Andromeda is no exception: at its heart lurks a black hole much larger than ours, weighing in at 30 million times the mass of the Sun (ours is only about 4 million). When the cores coalesce, the two monster black holes will merge, creating a single black hole with 34 million solar masses.* Even a 1 percent chance of getting dropped near such a monster is a little higher percentage than I'd like. Still, if we can manage to escape getting swallowed by the black hole, there's yet another problem: gas.

While there is not enough gas left over during and after the merger to form new stars, it takes far less gas falling into an SMBH to create an active galaxy. While not explicitly calculated by Cox and Loeb, it is implied in their models that some mass will drop toward the center of the merger, where it can form an accretion disk and be consumed by the black hole there. As you may recall, many active galaxies blasting out copious amounts of radiation and matter seem to have odd shapes, implying they recently suffered collisions.

If this were the case, then once again our galaxy—well, *Milkomeda*—will become active. Beams of matter and energy will blast out of the supermassive black hole in the core, and, if the Sun is in the wrong place at the wrong time . . . well, you know what happens to us: chapter 5 discussed these beams from a black hole. Now imagine them being a thousand times more powerful, with us in their path. If the Sun drops toward the core of the new galaxy and the supermassive black hole there decided to throw a fit, we're in for a very bad ride. However, if the Sun is ejected off to 100,000 light-years away from the core, then the odds of intersecting one of those beams is rather small . . . and the work of Cox and Loeb indicates we have a far better chance of heading out, not in.

Of course, we're talking about a time maybe five billion years from now. All politics is local, they say, and if we're still around we'll probably be contending with a star on its way to becoming a red giant and white dwarf. When your own small town's politics are so messed up, who has time to worry about the big-city slickers and what they're doing so far away?

---

* One prediction of Einstein's relativity is that merging black holes will actually cause a ripple in the fabric of space and time, like taking a bedsheet and frantically whipping it up and down. However, the *gravitational waves* resulting from two SMBHs merging are probably not strong enough to have any real effect on stars and other matter around them.

CHAPTER 9

The End of Everything

BLACK. NOTHING. EMPTY.

Everything is dark. No stars dot the inky sky, no galaxies can be seen.

They are all long since dead, gone, disintegrated as their very constituents have decayed into nothingness.

Nothing has occurred in the Universe for countless years. It is a cold, almost entirely empty void.

For trillions of trillions of years, this emptiness endures. But then, suddenly, in one tiny corner of the Universe no different from any other, a phase change snaps into existence. Like crystals growing in a saturated solution, this realignment in the very structure of space and time expands. It spreads outward at nearly the speed of light, enveloping more and more space.

What it leaves behind is . . . nothing. Or at least, nothing we can understand. Matter, energy, even space and time are destroyed, transformed in the wake of this quantum bubble.

When it is done, it has consumed the entire Universe. And what is left after that is something we may never know.

A NOTE ON EXPONENTIAL NOTATION

You may be familiar with what's called exponential or scientific notation: using exponents to represent very large or very small numbers. So instead of writing out 10,000,000,000, it's easier to refer to it as 10^{10}: a 1 followed by ten zeros. Similarly, very small numbers are written using a negative exponent: $10^{-7} = 0.0000001$ (the 1 is seven places to the right of the decimal point). This chapter is rife with scientific notation because the numbers involved will get staggeringly large very quickly. However, this introduces a slight prejudice in our barely evolved human brains that can trick even those of us familiar with the notation.

The number 10^{12} looks like it is only a little bit bigger than 10^{11}, but it's actually ten times larger (1 trillion versus 100 billion). Worse, 10^{20} *looks* only twice as large as 10^{10}, but it's actually *10 billion* times larger! Even for someone experienced in this notation, it can be difficult to appreciate at a glance. Right now, the Universe is a little over ten billion years old: 10^{10}. But far, far in the future, when it's 10^{20} years old, the length of time spanned by 10 billion years will be a tiny, tiny fraction of the total age of the Universe. Keep this in mind, because by the end, even 10^{20} will be an infinitesimal amount of time compared to the length of our journey.

DEEP TIME

So far, we have examined a series of singular events that wreak havoc on our little planet: exploding stars of different flavors, catastrophic impacts, the death of the Sun.

Certainly, these events are exciting, and of course it's the big flaming explosions that make headlines. A tree getting hit by lightning and burning to the ground might make the local newscast, but one that simply decays from within and falls over from rot after fifty years won't even get noticed.

But while we might never get hit by an asteroid or fried by a GRB, the Earth is aging. *Everything* is aging. Even if we manage to survive the death of the Sun, how do we survive the aging of *the Universe itself*?

The answer may be bleak indeed: *we don't.* While you've been reading this book, the Universe has aged. Maybe a week, maybe a few days if you're a fast reader, and during that time the Sun has eaten through a few trillion tons of hydrogen, stars have exploded, and the volume of the Universe has increased. We're all a little older, and so is the cosmos. After you're done with this book and you put it on a bookshelf, it will age. It will *always* age. It's inevitable: it will be a year older, a decade, a millennium. It will have decayed into dust by then, no doubt, but the atoms in that dust will age too. Someday they will be millions of years older, billions. *Trillions.*

And even that is a microscopic drop in the ocean of time. Time may very well go on forever, and a trillion years will be like the blink of an eye. The Universe will continue to age, and as it does, it will change. This change will be profound: it is more than just massive stars dying and galaxies colliding; the very nature of the Universe and the things in it will change fundamentally over stretches of time so long we don't even have names for them.

What will the Universe look like in a trillion trillion *trillion* years? How about a trillion times that age?

Different. It'll be different. But we'll never know: we won't be around to see it. And by that I mean nothing like humans at all, nothing like life as we know it. Not even matter as we know it will survive to see this stage of the cosmos.

Time and tide wait for no man. But deep time waits for *nothing.* Not even matter.

The only way to understand this look forward is to first take a look back, all the way back, to the beginning of the Universe. It may seem like a long time ago, but I promise you, soon it will seem like yesterday afternoon.

A VERY *BRIEF HISTORY OF THE UNIVERSE*

In the beginning, there was nothing.

Then there was everything.

DETAILS, DETAILS

Maybe that's a little *too* brief. The details turn out to be important.

How we understand the beginning of the Universe—and its fate—depends on how we see it now. By carefully examining the clues coming down from space through our telescopes, we actually know a surprising amount about what the Universe has been doing the past few billion years. Perhaps even more surprising is how much we can extrapolate about what it will do in the future . . . and so far into the future that billions of years will seem as but a whisper, a mere tick of the cosmic clock.

The events described below will seem like science fiction, I suppose, since things get pretty weird at very short times after the Universe formed, and in the mind-numbingly long stretches of future time. But as far as we can tell, this is science *fact,* based on solid evidence. Like any examples of conjecture, there may be pieces we are missing that affect what really happened, or what really will happen. That is the nature of science; more observations and more information always lead to a refinement of the results. Science asymptotically approaches reality, and it's hard to say just how far up the curve we are now.

But even with that caveat, the future of the cosmos is fascinating, if bleak. But as is the case in science and in stories, we have to start at the beginning.

Oh, and I better warn you: the very dawn and the very dusk of the Universe are times when things are completely different from what we see around us now. Be prepared to stretch your mind a bit.

The Universe is many, many things—it's literally everything—but it's also a damn odd place.

IN THE BEGINNING

Some 13.7 billion years (plus or minus 200 million years) ago, the Universe exploded into existence.

This mere statement causes a substantial amount of confusion. Astronomers refer to this event as the Big Bang, or, more accurately, we use the term *Big Bang* as a *model* for what we think happened. What's the distinction? Well, for one, the event was neither Big nor was it a Bang. When it popped into being, the Universe was smaller than the size of a proton, so it wasn't terribly big.* And there wasn't a bang either: it was more of a pop, or a snap.

By observing the Universe as it is now, we can run the clock backward and figure out what it was like in the past. What we discovered is that in the past, the Universe was hotter and more dense. The farther back in time you go, the hotter and denser it gets (the reverse is true as well: the older the Universe gets, the less dense and the cooler it gets). It gets smaller too: the Universe is expanding now (more on this in a moment), so in the past it was smaller. Eventually, you go back far enough to a time when the Universe was basically just a singular point: an infinitely small, infinitely hot, infinitely dense object.

Well, that's *weird*. And it's probably not even technically correct. As we turn the clock back, the Universe shrinks. At some point we see it as the size of a present-day galaxy, and then a star, and then a planet, and then a grapefruit, and then an atom. When it gets smaller than an atom, the weird world of quantum mechanics rears its head once again. One of the most fundamental rules of QM is that many characteristics of objects are related, and the more you know about one the less you know about another. The more carefully you measure an electron's position, for example, the less you know about its velocity. The more you

---

* On the other hand, you can argue that since the Universe is all there is, *everything* there is, then the explosion happened everywhere all at once, and so it *was* big. That's just semantics, though.

nail down one aspect of an object ruled by quantum mechanical processes, the slipperier another property becomes. It's almost as if there is a cosmic censorship going on, happening at very teeny-tiny scales. The closer you look, the fuzzier things get.

In practical terms—and I'm not sure what the word *practical* even means on scales like this!—this tells us that when the Universe was really, really small, there is very little we can know about what it was truly like. Our equations and understanding of physics tell us a great deal about the state of the Universe when it was a day old, an hour old, a second old . . . even a tiny fraction of a nanosecond old. But if you go back far enough in time, to when the Universe was literally 10^{-43} second old,* our physics breaks down. There is a veil hiding the true beginnings of the Universe, farther back than which we can never directly see.

That's another reason scientists prefer not to call this event the Big Bang. We don't know much at all about the event itself, whether it was a bang or not. We can only figure out what happened right after it.

Still, you might wonder what happened *before* the Big Bang. This is a natural question, and there are two ways to think of it. One of them is that the question is meaningless. That may sound like a cop-out, but let me ask you this: what's north of the north pole?

That question has no meaning, right? If you travel north, you get to the north pole, and you're done. There's no more north to go.†

Now remember that time itself was created in the Big Bang. Before then, there was no time, so there was no "before." The question has no meaning.

That's pretty strange, even for cosmology and quantum mechanics. It's also unsatisfying. We're used to things existing for a finite stretch of time, embedded in a bigger stretch of time. A symphonic concert might

---

* That means 0.0000000000000000000000000000000000000000001 second old.

† To paraphrase the great philosopher-scientist Nigel Tufnel from *This Is Spinal Tap:* "How much more north could it be? The answer is none. None more north."

start at 7:00 p.m. and end at 8:24. But there existed things before the symphony started (the orchestra arrived at the concert hall, warmed up, filed on stage), and things continue after (the brass section empties their spit valves, the members leave the stage, they go home and watch reruns of *Gilligan's Island*). So how can there be a beginning of time, a point on the timeline before which there is nothing?

There may be a way out of this conundrum. There are some theories that say that the Universe is not really all there is. There may be some sort of meta-Universe out there, some framework from which we are forever locked away, and our Universe is just a subset of it. This Universe existed before ours did and is much like ours, with the same or similar laws of physics controlling its behavior, including quantum mechanics. In the chapter on black holes we saw how particles can pop into existence spontaneously. There is a possibility that a tiny blip in the fabric of this other Universe's space-time suddenly came into being, something like the creation of particles ex nihilo. Under some conditions, this herniated region of space and time would quickly collapse, but it's also possible, in the realm of quantum mechanics, for this region to *grow.* Something like a black hole, it's disconnected from the greater Universe around it, and becomes its own entity, its own Universe. Space and time and energy and matter simply spring into existence inside. After a few billion years it expands to the point where stars form, galaxies take shape, and on a planet somewhere lost in all that volume of space, a person reading a book is scratching his head and thinking the book's author has lost his mind.*

At the moment, we don't know for sure whether there was anything before our Universe existed, or if that idea even has meaning at all. But since the Big Bang theory was first postulated, we've learned quite a bit about what happened after that first 10^{-43} second.

---

* Anything's possible.

A LITTLE AFTER THE BEGINNING: T + 10^{-43} SECOND TO TODAY

We know that the Universe went through several different phases of its life prior to the one we live in now. In the very early Universe, when it was still unfathomably hot and dense, it consisted of a stew of bizarre subatomic particles held in sway by unfamiliar forces. As the Universe expanded and cooled, different types of particles were able to form and become stable (whereas before it was just too hot for them to exist, like an ice cube won't last long on a frying pan). To make it easier on themselves, physicists divide the timeline of the Universe into different slices, different eras, based on what particles were present and what forces dominated at the time.

After just one microsecond (10^{-6} second), things had settled down enough for protons and neutrons to form from the thick soup of subatomic particles called *quarks.* After one second, one tick of the clock, was the period of *nucleosynthesis,** when conditions were similar to those in the core of a star. The heat and density allowed some protons and neutrons to come together and form stable nuclei. For about three minutes after the nucleosynthesis period started, the subatomic particles smashed into each other and created an entirely new kind of matter: helium (two protons plus two neutrons). Even a trace of lithium (three protons and three or four neutrons) was formed, but nothing heavier than that—the more complex reactions needed to make carbon and neon never got a chance to occur.

This left all the matter in the entire Universe divided into roughly 75 percent hydrogen, 25 percent helium, and a trace of lithium.

There was no calcium, no iron, no oxygen. For that matter (pun intended) there were no stars, no planets, no galaxies. Everything was pretty simple at this point, just a lot of extremely hot gas strewn into long filaments and ripples: fluctuations in the cosmic matter distri-

---

* Literally, the creation of new nuclei, new elements.

bution created by fluctuations in the explosion of the Big Bang itself.

These streamers would soon begin to collapse under their own gravity. Under conditions still not fully understood, the matter would form the first stars at about T + 400 million years. Galaxies themselves formed around the same time, collecting along the matter filaments, creating a vast spongelike network of fantastic galactic clusters streaming throughout the Universe.

And so, after 13.7 billion years, here we are.

HOW WE KNOW WHAT'S SO

All of this is probably a little overwhelming. It may even strike you as ridiculous! It's so far outside our comfort zone, our usual thought processes, that it might seem as if scientists are just making it all up.

I promise, we're not. There is a logical series of steps that lead to our understanding of the early Universe.

One of the very first people to think about the Universe as a whole was a German astronomer named Heinrich Olbers. In the early 1800s, when Olbers was studying the sky, it was assumed that the Universe was infinitely old, and that it was infinite in extent. There was no reason to think otherwise. But as Olbers realized, this raises a problem. If the Universe is infinite, and populated with stars throughout its extent, then no matter what direction you looked eventually you'd be seeing the surface of a star. No matter how teeny-tiny a section of the sky you chose, a line drawn from you out into space in that direction must hit a star at some point. It might be a bazillion light-years away, but if the Universe is indeed infinite that is but a mere stroll compared to that infinity.

And that's the problem. The apparent size of a star gets smaller as it gets more distant, of course, so it also appears fainter. But the drop in size and brightness is compensated by the Universe being literally infinite. The number of stars increases with distance, and in fact the number of stars *increases* at the same rate at which the brightness *decreases.* The two cancel out.* So if the sky appears full of stars, literally, with no apparent space whatsoever between them, then the entire sky will glow with the same brightness as a star itself. To any observer inside such a Universe, it would be as if the sky were as bright as the Sun, *everywhere you look.*

Obviously, such a Universe would be uninhabitable. Also just as obviously, our Universe doesn't behave that way.

This is what Olbers pointed out, and the conundrum is now known as Olbers's paradox. This problem baffled people for some time, and the answer to the paradox came from a somewhat surprising source: Edgar Allan Poe.

Yes, *that* Poe. Besides writing scary stories and depressing poems like "The Raven," he was quite a deep thinker. It occurred to him that perhaps the problem lay not in the Universe, but in our underlying assumption: what if the Universe were *not* infinite in space and/or time? If the Universe were finite in space, then you'd run out of stars at some distance from Earth. And if it were finite in time—that is, it had a beginning—then there simply hasn't been enough time for the light from very distant stars to reach us. Paradox solved.

In fact, Poe was right. In his 1848 work *Eureka,* he wrote:

> Were the succession of stars endless, then the background of the sky would present us an uniform luminosity, like that displayed by the galaxy—since there would be no point, in all that background, at which would not exist a star. The only mode, therefore, in which, under such a state of affairs, we could comprehend the

---

* A little math: Like gravity, the brightness of a star decreases with the square of the distance to the star—double the distance to a star and it will appear one-quarter as bright. But if stars are distributed evenly throughout the Universe, you're basically adding up all the light from stars at a given distance from you, and they form the surface of a sphere. The area of the surface of a sphere depends on the square of its radius. So brightness drops with the distance squared, and the number of stars goes up with the distance squared—canceling each other out.

> voids which our telescopes find in innumerable directions, would be by supposing the distance of the invisible background so immense that no ray from it has yet been able to reach us at all. That this may be so, who shall venture to deny? I maintain, simply, that we have not even the shadow of a reason for believing that it is so.

This was radical thinking for its time. While it was common in mid- to late nineteenth-century society to assume that the Universe had a beginning because it said so in the Bible, this was somewhat unsatisfying to a scientist. Poe changed that.

Less than a century later, the astronomer Edwin Hubble, together with other astronomers such as Vesto Slipher and Ellery Hale, made one of the most shocking discoveries in the history of science: essentially every galaxy they could observe appeared to be rushing away from us. This was so difficult to believe that it took years before there were enough observations to convince everyone, but the evidence was undeniable: the Universe itself is expanding.

This had profound implications. If galaxies were moving away from us, then they grew more distant as time went on. That in turn means they were all closer together in the past. If you run the cosmic clock backward far enough, then, at some point in the past, every galaxy, *every bit of matter and energy in the Universe, would have all been in the same spot.*

This meant the Universe had a beginning, a point in time when it all began. Matter and energy rushed outward from that point in time, expanding evermore. Albert Einstein had already been working out the general equations that govern the behavior of time and space when Hubble and his team discovered the cosmic expansion, and the news of the discovery electrified him. It was soon accepted by scientists that Einstein's work was correct, and that the Universe itself could be described using mathematics.

Thus the Big Bang model was formed.

Over the years, the model has been reworked, refined, with parts added and others taken away. When an astronomer uses the term *Big Bang,* she doesn't just mean that singular point in time 13.7 billion years ago; she is also implying a vast amount of other work done to make the model fit what is observed about the Universe. And, in fact, it is one of the most successful scientific theories of all time.*

One critical factor in confirming the Big Bang model is the finite speed of light. That may sound weird, but it's this finite speed that allows us to see what the Universe was doing in the past. Imagine that the speed of light were infinitely fast. If we looked at a galaxy 10 billion light-years away, we'd see it as it is *right now,* at this very moment. It would probably look a lot like ours, and there's not a whole lot about the Universe we could learn from it.

Instead, though, we have a wonderful characteristic of the Universe: light is not infinitely fast. It's *pretty* fast, covering 186,000 miles every second (about a foot per nanosecond, if that helps you any), but the Universe is so big it takes a long time for that beam of light to make it here from some distant galaxy.

What this means is that we don't see galaxies as they are right now; we see them as they were when they were younger. Telescopes are very much like time machines in this regard—the farther away we look in space, the farther back we look in time. How do we find out what the Universe was like five billion years ago? Easy: find galaxies that are five billion light-years away and take a look.

And why stop there? Our telescopes are huge, and our detectors sensitive. We have seen galaxies well over 12 billion light-years away, so we're seeing them as they were when the Universe itself was about a billion years old. Because of this, we can actually see what galaxies looked like when they were young, and discover what happens when they age.

---

* I am using the word *theory* as a scientist means it: a set of ideas so well established by observations and physical models that it is essentially indistinguishable from fact. This is different from the colloquial use that means "guess." To a scientist, you can bet your life on a theory. Remember, gravity is "just a theory" too.

We can also detect and analyze the gas that lies between galaxies in the distant Universe, which in turn tells us even more about early conditions. In fact, radio telescopes tuned to the microwave part of the spectrum have detected a uniform hiss coming from all over the sky. This hiss is not noise: in a very real way it's the cooled light from the fireball of the birth of the Universe. After about 100,000 years, the Universe had expanded and cooled enough that the matter became transparent to light, meaning that light could travel freely through it. Before then, a photon wouldn't get very far before being absorbed by some bit of matter. This light, free to move across space, has since "cooled" as the Universe expanded, and has been able to travel to our waiting instruments.

These characteristics—and many, many more—have provided to us a wonderful series of clues on the way the Universe behaves. Because of this, we have a fairly good grasp on what the Universe was like almost all the way back to its birth, nearly 14 billion years ago.

But what about its future? Is it possible to take what we know about physics and astronomy and extrapolate the eventual fate of the cosmos?

Yes, it is. We can get a fairly good idea of what the Universe will be like in the next few billion years (for example, our local neighborhood will look surprisingly different pretty quickly). As we look farther into the future, though, our crystal ball gets cloudier, but given what we see and know we can guess in broad terms what will happen.

I'll give it to you straight: things don't look good for us. If we want to survive into the far future—and I mean *far*—we'll have to change ourselves in such fundamental ways that I wouldn't even consider the result to be human anymore. And even then, escape from the Universe's eventual demise may be impossible.

And yet there is still hope. Maybe not for *us,* exactly, but for whoever comes next. Maybe there won't be anyone next, but the Universe may yet get another chance to try.

THE AGES OF THE UNIVERSE

As the Universe ages, it changes profoundly. The time scales of these overall changes themselves change, getting longer as the Universe ages. When the Universe was young it changed rapidly. For example, one of the first big changes occurred just 10^{-35} second after it was created. Before this time, all the forces of the Universe—gravity, electromagnetism, and the nuclear forces—were combined in one unified force, and in fact this is called the Grand Unification Epoch. Today, these forces are as different as they can be, but when the Universe was incredibly hot and dense, they were indistinguishable. But just after that razor's slice of time following the cosmic birth, the temperature and density dropped enough that the forces started acting differently.

It's an incredibly short length of time, 10^{-35} second. But it was enough for the Universe to change profoundly. It went through many such changes: dropping in temperature and density such that particles like protons and electrons could form, and then dropping again such that these could come together to make more complicated elements, and then form stars, galaxies, and finally us.

There are lots of ways to divide up the Cosmic Epochs. A good way is to look at what's producing the most amount of energy at that time. Right now, that would be stars. However, the stars will all eventually die out, and the current age will end. What then?

Astronomers Fred Adams and Greg Laughlin looked into this idea in great depth in their book *The Five Ages of the Universe*. As the title suggests, they found five ways to divvy up time in the Universe. Up until stars formed was the Primordial Era, which we just toured above. The current era of stars they dubbed the Stelliferous Era. After that is the Degenerate Era, then the Black Hole Era, and finally—forbiddingly—the Dark Era. The time covered by these eras is staggering, and difficult to grasp. While reading their book I had to constantly take a step back and laugh at the numbers. Maybe that was a defense mechanism on my part, like whistling past a graveyard.

Actually, that analogy is a little more on the mark than I'd like.

That's just a gentle warning. We're about to take the longest journey you've ever been on. It'll last so long that even using scientific notation gets overwhelming. You'd better sit back and relax. You're going to be reading this chapter for a long, *long* time.

THE STELLIFEROUS ERA: T + 10^8–10^{15} YEARS

The era of stars began with the birth of the first stars. It's not known precisely when that happened, but the best estimates put it at about 400 million years after the birth of the Universe. Theoretical models show that it wasn't until roughly then that the gas distributed throughout the Universe was cool and dense enough to collapse under its own gravity.

Observational evidence has mounted for this date as well. Although we have never directly detected these pioneer stars—they would be so distant now that directly observing them would be nearly impossible—they had an effect on their environment, and that *can* be detected. These stars would have been made entirely of hydrogen and helium (and again, a trace of lithium), making them relatively simple compared to modern stars. Such a chemical composition made it possible for the early stars to be much more massive on average than current ones (the heavier elements in modern stars make them hotter, so they "switch on" at a much lower mass). Some models put these stars at well over 100 times the mass of the Sun. They flooded space with ultraviolet light, which ionized the hydrogen atoms around them, tearing the electrons off.

These electrons *polarized* the light from the stars: in effect, this means the waves of light coming from the stars were all aligned, like people in a room all facing the same direction.* This polarization effect can still be detected today, and the observations agree with the theoretical models on the time when stars first appeared.

---

* Light reflecting off water and metal can get polarized as well. Sunglasses that are polarized can block just the type of light waves that are aligned in that way, greatly reducing the glare of light reflected off cars and puddles.

Also, at the ends of their short lives, these stars would have exploded as massive supernovae, scattering the Universe's first heavy elements into the surrounding environs, from which the next generation of stars would form. The first stars probably created gamma-ray bursts when they exploded; these might yet be detected too.

We still live in the Stelliferous Era. Stars are the dominant feature of the Universe, and produce most of the energy we detect. As we saw in the last chapter, the available source of gas in the Milky Way to make stars will run out in the next few billion years, although some galaxies may use up their gas more slowly. But one way or another, the gas will eventually run out, and essentially no more stars will be born anywhere in the Universe.*

We know the Sun will last as a normal star for several billion more years before turning into a red giant, frying the Earth, losing its outer envelope, and then "retiring" as a white dwarf (chapter 7). But the length of time a star lives depends almost entirely on its mass. A star with a lot more mass than the Sun eats through its fuel far faster, and may only live a few million to a billion years. However, stars with *less* mass will live longer.

The lowest-mass star that can currently exist has about 0.08 times the mass of the Sun. Below that limit, the core isn't hot enough or under enough pressure to fuse hydrogen into helium. This type of star is small (one-tenth the Sun's diameter), dim (one one-thousandth the Sun's luminosity), cool (with a temperature of about 5,000 degrees Fahrenheit), and red. Not surprisingly, these stars are called *red dwarfs.*

Imagine you take a large rock and hit it with a sledgehammer, shattering it. If you look at the pieces that remain you might see a few large pieces, a few more that are smaller, and a lot of little pebbles and shards.

---

* Gas does get recycled in galaxies: stars explode, other stars lose mass in a stellar wind, and so on. It's possible in some galaxies that stellar birth may continue for as long as another trillion years, but those are the exceptions, not the rules. In a trillion years or so, star formation will effectively cease.

That is a natural size distribution in stars as well: when a cloud collapses and forms stars, only a few will be really big, some will be smaller, and more smaller yet. The vast majority will be the smallest type; it's estimated that 75 percent of all the stars in the Universe are red dwarfs.

Although they have a small fraction of the mass of the Sun, red dwarfs are incredibly miserly with their fuel and can last far, far longer. A very low-mass dwarf can reasonably expect to shine for the next several *trillion* years.

This is longer than any other kind of star in the Universe. If we let the cosmic clock run forward, we see the last stars being born in a few hundred billion years. Very rapidly after that all the massive stars will be gone, since they don't live long. The last core-collapse supernova in the Universe may occur only a hundred million years after the last massive star is born. This is a tick of the clock compared to how much time has elapsed in the Universe at that point.

Sometime not long after that, somewhere in the Universe, a star just barely too low-mass to explode will age and die, expanding into a red giant, blowing off its outer layers, and fading away as a white dwarf. It is part of a long, long line of such events: there are 100 billion galaxies in our Universe, each with an average of about 100 billion stars.

As time goes on, trillions of stars with lower and lower mass fade away and die. Stars with the lowest mass will take the longest, but they'll all cross the finish line at some point.

If we wait a sufficiently long time—oh, say, a trillion years—all stars like the Sun will be long gone, and only the lowest-mass dwarfs will remain. You might think galaxies would be dim and red then, only illuminated by the tiny stars. Interestingly, though, galaxies may be as bright at that time in the distant future as they are today. We saw in chapter 7 that the Sun steadily increases in brightness as it ages. All stars do this, even red dwarfs. Calculations done by *The Five Ages of the Universe* authors Adams and Laughlin, together with their colleague Genevieve Graves, indicate that a star with one-tenth the Sun's mass will live for about 10 trillion years. As it ages, it gets brighter and slightly hotter. What they found in their models, after adding up all the light from all the stars in the galaxy and then letting the galaxy age, is that the amount of light given off in toto by dwarfs increases roughly as quickly as light from the more massive stars fades as they die. In other words, the total light emitted by a galaxy will stay roughly constant for several hundred billion years, with the ever-brightening dwarfs picking up the slack as massive stars die off.

As the red dwarfs heat up, they will change color too. A hotter star gets bluer, and so too will red dwarfs. It's possible that for a few dozen billion years the galaxy will shine with a demonic red hue, and then this will slowly morph to a vibrant blue.

But all good things . . . , as they say. Even dwarf stars eventually die. Unlike the Sun, which can only fuse fuel in its core, the smallest red dwarfs circulate their fuel. Like hot air rising and cool air sinking,* the helium created in the core circulates upward and mixes with the rest of the star. As the hydrogen falls into the core it can fuse, forming more helium, which then mixes more with the star.

Eventually, the star runs out of hydrogen—and unlike the Sun, which just runs out of *available* hydrogen in its core, the dwarf *totally* runs out. Gone. Kaput. All that is left in the star is helium, and it lacks the mass to fuse it into carbon. The star cools, the helium contracts, and it becomes a pure helium degenerate white dwarf (see chapter 7 for details on this odd quantum state).

In seven or eight trillion years' time, in the Milky Way (well, Milkomeda, after we collide with the Andromeda galaxy, and probably consume all the smaller galaxies in the Local Group as well) the last dwarf star will become a white dwarf. For trillions of years the galaxy will have glowed a beautiful blue, but that too shall pass.

Interestingly, in this late stage of the Stelliferous Era, some even lower-mass stars will still be able to shine. Because high-mass stars

---

* In fact, it's *exactly* like that: gas in a dwarf circulates in precisely this manner.

create heavier elements like iron and magnesium, stars forming later get imbued with these materials. Heavier elements make a star hotter (they absorb the light from the star, trapping the heat in), so lower-mass stars—perhaps even as lightweight as 0.04 solar mass—will be able to get fusion started in their cores. But again, we have to consider the span of time: even if these stars stave off turning into white dwarfs for 15 trillion years, *that time will still eventually come.* At some point, all stars in the Universe will be gone, having become white dwarfs, neutron stars, or black holes.

The tiny white dwarfs fade with time (neutron stars cool even faster). Eventually, the galaxy contains no stars actively fusing elements in their core at all. Over the next few trillion years these stars fade too. By the time the Universe is 100 trillion years old, the galaxies—and therefore the Universe itself—will be dark.

LOST HORIZON

In the distant future, not only will the Universe be darker, it'll look a whole lot emptier as well.

Standing on a beach looking toward the horizon, you can see only so far out. The Earth curves downward, hiding more distant objects from view. The visible horizon is only a few miles away, and you cannot see things any more distant.

The Universe has a horizon too. Since it's 13.7 billion years old, we cannot see any objects more than 13.7 billion light-years away. The Universe might be bigger than that, but the light emitted by any objects farther away than that distance has not had enough time to reach us, so we don't see them.

In fact, it's actually worse than that. The Universe is expanding; the fabric of space is literally being stretched out. Objects farther away appear to be receding from us at greater speeds. If you look out to a great enough distance, galaxies appear to be receding from us at the speed of light. We cannot detect such galaxies: their light is approaching us at the same speed that space is expanding. Like running on a treadmill, that light can't get anywhere, so it never reaches us.*

And it's even worse than that. In 1998, it was discovered that not only is the Universe expanding, but its expansion is *accelerating.* Not only is the Universe getting bigger, the *amount* it's getting bigger is bigger every day.

This has a rather depressing outcome for the distant future. Because the Universe is accelerating, galaxies that are currently inside our horizon (because they are receding from us at less than the speed of light) will eventually move outside our horizon (because they will accelerate relative to us to beyond the speed of light).† This means that over time distant galaxies will fade as the expanding Universe sweeps them out of our view. As time creeps ever forward, galaxies that are closer to us now will slip away, and the cosmic horizon will close around us like a noose.

However, it won't tighten too much. Space expands, but this expansion can be counteracted by gravity. You might expect that, say, two stars orbiting each other will get farther apart as space expands between them. However, that's not the case. Since the two objects have gravity, and they are bound to each other—that is, their gravity holds them together—*space doesn't expand between them.* It's just another peculiar outcome of relativity and the way space-time behaves.

This means that even though the Universe is expanding, and even accelerating, the cosmic horizon won't continue to shrink forever. The Local Group of galaxies—the Milky Way, Andromeda, and a dozen or two smaller galaxies—are gravitationally bound to each other. We know that by 10 billion years from now we will have merged with Andromeda, and over time we'll gobble up all the remaining smaller galaxies as well.

---

* In reality, the explanation of this is far more complicated and involves invoking Einstein's theory of relativity. I'll spare you that and just leave you with the treadmill analogy, which is close enough.

† Remember, as discussed earlier, that space can expand faster than the speed of light. The distant galaxies aren't really moving faster than light; the space in between the galaxy and us is expanding such that it *appears* the galaxy is moving faster than light. Think of it as the track on the treadmill stretching as you're running on it.

Some calculations show that the cosmic horizon will shrink down to encompass just the Local Group's volume of space in something like 100 billion years, still during the Stelliferous Era. By that time, the Local Group will be one giant elliptical galaxy.* From our point of view, we will see closer galaxy groups like the Virgo Cluster fall over the horizon, but our own will always be visible.

Eventually, our view will be extremely limited: as far as we'll be able to tell, the entire Universe will consist of our one massive galaxy with literally nothing outside of it. Any new species that evolves during this time will have no clue that, once upon a time, a Universe far vaster than their own once teemed with galaxies and stars.

What will their cosmology be like?

The tightening of the horizon will occur far sooner than all the stars in the galaxy will die out, hundreds of billions of years compared with tens of trillions. Still, it'll be a harbinger of things to come, of a Universe growing increasingly darker.

It should be noted that the acceleration of the cosmic expansion does mean one thing for sure: the Universe will not recollapse. Before the acceleration was discovered, it was still a matter of some debate whether the Universe would expand forever or whether the combined gravity of all the matter in it would slow, stop, and eventually reverse the expansion. But the discovery of the acceleration pretty much put an end to that debate. The Universe will expand forever, ever faster, while (somewhat ironically) our view of it will get smaller and smaller, until we have our own private Universe just a few million light-years across.†

---

* I'm at a loss for a name for this galaxy . . . MilkLocalGroupeda? Androgroupyway?

† There are some theories stating that, depending on what's driving the acceleration, the Universal expansion may overwhelm gravity. Eventually, all of space will stretch, including space in between bound gravitational objects. If that's the case, then the horizon will continue to move in while space stretches. Eventually, everything will stretch—galaxies, stars, planets, even atoms. At some point, everything will get torn asunder as space itself shreds apart. For some reason, scientists call this idea the Big Rip. This turn of events seems pretty unlikely given what we know about the Universe, but it's something to consider.

THE FAULT LIES IN THE STARS

What does this mean for us, for humans? To a good approximation, it means that we have about 100 trillion years to get our affairs in order. After that, we won't have enough light to read our books by. Things'll get boring.

Assuming that anything resembling humans still exists a thousand times the current age of the Universe from now, there *are* ways to extend the stars' reign. Physically colliding stars—literally smacking them into each other—to make new ones will help. But how long can you do that? If you decide you need a star like the Sun, you can smush together a few dwarfs and get a star that shines for another few billion years. Remember, though, that the Universe is *trillions* of years old by this point. A billion years is a pittance in comparison. When the Universe is 100 trillion years old, our descendants will be out of fuel, out of stars, and out of luck.

The time scales here are forbidding. When we reach this point in the age of the Universe, galaxies will have lived the vast majority of their lifetime populated only by dwarfs. Think of it this way: currently, our galaxy has only been around a tiny fraction of its potential life span. Right now, as you read this, despite the Universe being over 13 billion years old, *99.9* percent of the galaxy's life still lies ahead of it.

We think of the Universe as being relatively unchanging, but in fact we live in a very special epoch compared with the dim future. By the time the last dwarf fades away, the galaxy will look back at the time stars like ours could exist in the same way you look back at the time you were a month old.

And after all that, we're only just getting started. We're about to enter a realm where even 100 trillion years is a single breath of time.

THE DEGENERATE ERA: $T + 10^{15}$–10^{40} YEARS

When the last normal, fusing stars die, the only objects left in the Universe that can generate energy will be white dwarfs, neutron stars, black holes, and degenerate low-mass objects that lack the capability to fuse hydrogen in the first place, called *brown dwarfs.** Because the Universe is dominated by these objects, this time period is called the Degenerate Era.

In visible light, the Universe will be pretty dark at this point. However, it won't be completely dark, since there *will* be a few scant sources of light.

White dwarfs will fade; when they are at about 10,000 degrees Fahrenheit they will shine with the same color as the Sun, getting redder as they age. When they reach a temperature of about 800 degrees Fahrenheit they radiate mostly in the infrared and will become invisible.

Every now and again a black hole may pass close enough to a white dwarf, neutron star, or brown dwarf to shred it and consume the debris. An accretion disk will form and shine brightly, but only as long as the black hole eats. Once the meal is gone, the light source shuts off (this may provide a temporary source of energy for any future beings looking to stay alive, but it really is only a short-term solution).

Brown dwarfs will have their moments as well. These failed stars give off visible light for a short time after they form because of their internal heat, but their lack of core fusion means they have no ongoing source of energy. Eventually they cool and glow faintly in the infrared.

But they still can get a second chance. Collisions between stars are incredibly rare in the present-day Universe because stars are so small compared to the distances between them. However, the word *rare* has less meaning as time stretches on. Something that has an incredibly small chance of happening in 13.7 billion years may become a virtual certainty over 100 trillion.

The Degenerate Era will actually last much longer than this, in fact, so collisions between stars will happen frequently once you grasp that

---

* There will still be planets; they will orbit white dwarfs and brown dwarfs, and probably many more will wander the Universe after being ejected from their home stars during planetary formation. However, planets don't generate energy, so they aren't of much interest to us here. They'll be frozen solid.

time scale. When two brown dwarfs merge, their mass will be just above the fusion limit, so a relatively normal star could result. In fact, if the collision is a little bit off-center, then matter from the two objects could be stripped off, forming a disk around them. It's entirely possible that planets could form from this material; is it too hard to imagine life forming under such circumstances? Their view of the Universe would be far, far different from ours. Their skies would be entirely dark except for the one sun burning during the day. No stars, no galaxies, no ribbon of milky gas streaming across the sky. What myths and legends would arise on such a planet?*

At any one time, perhaps a hundred or so of these odd stars will shine in a galaxy. But again, these new stars would shine briefly, then suffer the same fate as the Sun did all those forbidding trillions of years in the past.

There will be other, brief flashes of light. A collision between two white dwarfs could result in an object whose mass is so high that it collapses into a neutron star or even a black hole. A Type I supernova may result, which to any denizens of this future era would be even more blindingly bright than to us: there will be literally nothing against which to compare it.

It's also possible that two low-mass white dwarfs could merge to form an odd type of "normal" star, much as the colliding brown dwarfs will, but again, this is a short-lived object (a mere few billion years!) and will fade with time.

If two neutron stars collide, then they will form a black hole with a gamma-ray burst to announce the merger (see chapter 4). But this fades within days, and the black hole itself will be dark, one among many millions of others orbiting a dark galaxy.

---

* Binary brown dwarfs are common: two brown dwarfs in orbit around each other. Owing to some weird effects of Einstein's relativity, the orbits of the objects will slowly decay with time. For a typical pair, the two objects will collide after about 10^{19} years, which is in the Degenerate Era. A merging of two brown dwarfs this way will almost certainly create a disk of material around them in the same way as an off-center collision would. This may be a "common" event during this era.

And it will get even darker. Time piles up. After trillions, quadrillions, *quintillions* of years, even the brown dwarfs go away. They merge to form normal stars that eventually die, or they get ejected from the galaxy entirely. In fact, after this length of time, the galaxy will have a hard time holding itself together. In the far distant future, the galaxy itself will evaporate.

GALACTIC PERCOLATION

Stellar collisions* are the culprit for this next stage of galactic evolution. A moving object has energy, and this energy can be transferred to another object (which allows us to do things like play pool, throw a baseball, hold a book, and so on). When two stars pass close to each other, they can exchange energy by interacting gravitationally. In general, what happens to two stars as they pass each other depends on their mass (it also depends on the sizes, shapes, and directions of their orbits, but we're being general here). The higher-mass object gives away some of its orbital energy to the lower-mass object. An orbit with lower energy is smaller, so the higher-mass star will sink closer to the center of the galaxy, while the lower-mass star will move outward. Over many such encounters, lower-mass stars "evaporate" away; they get ejected from the galaxy to wander the depths of intergalactic space.

The higher-mass stars drop to the center of the galaxy, where an unpleasant host awaits them: a supermassive black hole (see chapter 8). Eventually, all the higher-mass stars in the galaxy will get eaten by the black hole.†

This process has been seen on much smaller scales: globular

* Astronomers use the term *collision* to mean any encounter where two or more objects interact with each other through gravity, and not necessarily to mean direct physical contact like a car crash.

† Unlike smaller black holes, which will tear a star apart because of tidal forces, a supermassive black hole's tides are far, far smaller, so stars will get eaten whole. There will be no accretion disk, and so no light emitted by the consumption of a star. Eating gas clouds, on the other hand, still will cause the black hole cosmic indigestion.

clusters—gravitationally bound spherical collections of roughly a million stars—are packed tightly enough with stars that collisions of this sort are more frequent. In all globular clusters, even after only a few billion years, the more massive stars tend to be closer to the cluster center, with lighter stars farther out.

The time scale for galactic evaporation is about 10^{19} to 10^{20} years (10 quintillion to 100 quintillion years), making this process currently undetectable in galaxies.

But the Universe is still young. Patience.

Incidentally, over this length of time, the odds of a star getting extremely close to the Sun go up, and close in on 100 percent. Even by the beginning of the Degenerate Era (T + 10^{15} years), it's likely another star will have passed close enough to the Sun to dislodge the Earth from its orbit and eject it from the solar system (of course, any star that passes that close is likely to eject the outer planets as well—and by this time, Mercury and Venus will have been swallowed by the red-giant Sun, so *Earth* will be the innermost planet). Given enough time, planets even closer to their stars will go; even by halfway through this era, it's very unlikely that *any* planet orbiting *any* star *anywhere* will not have been ejected from its system. By the time the galaxy itself has evaporated through stellar collisions, there may be ten times as many planets as stars roaming intergalactic space, frozen to their cores and utterly uninhabitable.

And they won't last forever anyway.

PROTON DECAY

By 10^{20} years after the Universe formed, galaxies will be dark and mostly dispersed. Black holes, neutron stars, white dwarfs, and brown dwarfs will roam the Universe (such as we can still see of it, owing to the smaller cosmic horizon), and illumination will drop to a feeble whisper of what it once was.

But even this ignominy is not quite the end.

Matter, it turns out, may not last forever. We already know that many types of atomic nuclei and subatomic particles decay. Uranium is radioactive: over time, a uranium nucleus will spontaneously split apart into lighter elements (a process called *fission*), and give off a tiny bit of energy. The time for any given nucleus to fission is random, but if you take a whole pile of them and take data as they decay, statistically you start to see trends. You can measure how long it takes half the sample to decay, for example, and that number is pretty consistent. For one kind of uranium, it takes 4.5 billion years for half the sample to decay and become lead. This length of time is then uranium's *half-life.* If you start with a pound of uranium, you'll have half a pound in 4.5 billion years, and the other half will be lead. Wait another 4.5 billion years and half the remaining uranium will turn to lead (leaving you with a quarter pound of uranium). In another 4.5 billion years you'll have an eighth of a pound. And so on. Eventually, it will all turn to lead, but you have to be patient.

Individual particles like neutrons decay too, in this case with a half-life of about eleven minutes. This only happens if they are alone, free to roam space; in a nucleus neutrons are stable (they like the company, one supposes). But when they decay, they create a little shower of smaller particles and energy.

Until recently, protons were thought to be stable forever. But "forever" takes on a different meaning when dealing with the time scales of the death of the Universe.

Protons are theorized to decay into lower-mass particles extremely rarely, on average after about 10^{33} to 10^{45} years (the exact number is unknown, so for argument's sake we can pick an intermediate time of 10^{37} years). Currently, no protons have been unequivocally seen to decay,* but scientists are fairly sure they will. Given time.

Time is all we have here. In a given sample of protons—like, say, a

* We don't have to wait that long to see one decay: if we collect enough together, say 10^{37} of them, then we should see one decay every year. This has been attempted, and still no protons have decayed while scientists watched. If the decay time is off by a little bit—say it's 10^{38}—then this makes the process more difficult to detect . . . but 10^{38} years is still small compared to the time we're talking about in this chapter.

white dwarf—half the protons will decay in 10^{37} years. In another 10^{37} years, half more will disintegrate, and so on. After a few times 10^{38} years or so they'll all be gone.

Like any other subatomic decay reaction, when a proton decays, it creates smaller particles and energy. By this time, almost all protons will exist inside other objects—white dwarfs, brown dwarfs, neutron stars. When they decay, the net result is that energy is released, heating up the object a bit.

So long after the last light of fusion has burned out, long after all the material objects in space have cooled to nearly absolute zero, we find another source of energy: heating from proton decay.

It's feeble, to be sure. Very, *very* feeble: in a given white dwarf, the energy released by proton decay is only about 400 Watts. My microwave oven needs more power than that! In fact, the entire galaxy, even if full of such decay-powered objects, will only shine with less than a trillionth of the power with which the Sun shines now. Worse, the light it emits will be incredibly low-energy, well into the radio range of the electromagnetic spectrum.

If we were to make a leap of faith (and this isn't a leap, it's a transgalactic hyperspace jump) and assume that some form of life is still around deep into the Degenerate Era, then they had better figure out a way to go green. The amount of energy available to them will be incredibly small. They won't even be able to make a bowl of popcorn.*

And they'll run out of time too. Every time a proton decays inside a white dwarf or a brown dwarf, the star loses that much mass. It's not much each time—protons are pretty small—but time has a way of adding up 10^{37} years in the future. White dwarfs will lose mass† and will eventually evaporate entirely. As they lose mass they go through some

* Or whatever non-proton-based food they eat while watching movies.

† Because of the bizarre nature of degenerate matter, lower-mass objects actually increase in size when they lose mass, the opposite of what we expect. White dwarfs start out roughly the size of the Earth, but in 10^{39} years or more, they'll actually expand to be as big as Jupiter.

weird stages. When they have roughly the mass of Jupiter, for example, they will have the same density as water (when white dwarfs first form they are millions of times denser) and will be made almost entirely of hydrogen; all the more complex elements will have fallen apart as their protons decayed. The temperature of the object will be so low that it will be frozen, a ball of hydrogen ice 100,000 miles across.

Eventually, this too will go away as the protons inside it disappear.

Even neutron stars will undergo this evaporative process. Having more protons inside, they'll take longer than white dwarfs to disappear. They'll be warmer too: they'll shine at −454 degrees Fahrenheit. Today that's considered extremely cold, but in the year 10^{38} they'll be the hottest objects in existence.

And they, too, shall pass.

Eventually, they'll lose mass through proton decay as well. At some point, their gravity will decrease enough that neutron degeneracy cannot be maintained, and the star will suddenly expand into something like a white dwarf. This won't help it, though; we know what happens from there.

By the end of the Degenerate Era, an incredible 10^{40} years in the future, all the galaxies will not only be dead, but their corpses desecrated. There won't be a single proton left anywhere in the Universe. There will be no more stars of any kind at all. No white dwarfs, no neutron stars . . . not even planets, which will have evaporated long before the white dwarfs did.

All that will remain are extremely low-energy photons, a few subatomic particles that don't decay (electrons, positrons, neutrinos) . . . and black holes.

THE BLACK HOLE ERA: T + 10^{40}–10^{92} YEARS

Black holes survive the Degenerate Era because of one simple reason: they aren't made of matter.

Chapter 5 covers black holes in detail, but basically a black hole is an object that is so dense that its escape velocity is equal to or exceeds the speed of light. Once a black hole forms, no information can come out of it, and it's essentially cut off from the Universe. Any matter it was once made of, or any matter that falls in, is *gone*. Since there are no protons, there is nothing to decay. They therefore persist.

At the end of the Degenerate Era, all that's left are black holes and an extraordinarily thin soup of radiation and subatomic particles. After 10^{40} years, we have entered the Black Hole Era.

Black holes can have masses as low as three times that of the Sun and as large as the monster supermassive black holes in the centers of galaxies that, in our current era, contain from a million to a billion solar masses.

During the time of galactic evaporation in the Degenerate Era, a curious thing happens. Out in the suburbs of the galaxy, black holes will be the most massive objects that still exist. Normal stars today can have far more than three solar masses—the most massive have about 130 solar masses or a tad more—but they will have long since exploded. The only objects left in the Degenerate Era are neutron stars (top mass: 2.8 solar masses), white dwarfs (top mass: 1.4 solar masses), and the far less massive brown dwarfs. Since the most massive objects tend to sink and the lighter ones float away in the evaporation process, after the process is complete the galaxy will really consist of (1) a single, central supermassive black hole that has eaten many of the smaller stellar mass black holes that dropped into it, (2) quite a few (perhaps millions) of stellar mass black holes that have dropped down toward the center of the galaxy but have not (yet) been consumed, and (3) a bunch of lower-mass objects at large distances, many of which will have physically left the galaxy completely.

During the galactic evaporation process, the central black hole may have consumed 1 percent to 10 percent of the galaxy's mass in all. So, for a galaxy that started off with a hundred billion stars, the black hole at the core will end up with a billion or two solar masses by the end of the Degenerate Era.

Not all galaxies live alone, though. As pointed out, the cosmic horizon will shrink, but only to the point where gravity offsets it. Some galaxies exist in clusters like the Local Group, but far larger. The Virgo Cluster is the nearest galaxy cluster, and it has perhaps two thousand galaxies gravitationally bound to it. In a process similar to the evaporation of a single galaxy, the Virgo Cluster will evaporate as well, given enough time. When it's all done, the cluster will consist of a single galaxy with a mass of about 10 trillion times the mass of the Sun. Eventually that MonoVirgo galaxy will evaporate, and the black hole in its core will have a mass of a hundred billion times the Sun, or possibly more.

However, because our horizon will be so close, we'll never be able to observe that black hole. We're stuck with our one-billion-solar-mass hole in the center of our galaxy. And you'd think that would be that. Once a black hole, always a black hole.

Well . . . *almost* always.

Also as discussed in chapter 5, black holes too can evaporate. The process is called Hawking radiation, after the physicist Stephen Hawking, who first postulated it. Although it is still theoretical—we don't have any black holes handy on which to test it—it's grounded in well-understood physics. The basic principle is that black holes can radiate away their mass in the form of subatomic particles because of weird quantum effects. The process is in general excruciatingly slow, and it goes even slower the more massive a black hole is.

Once again, though, we have to be careful when we talk about "slow." When we have ten thousand trillion trillion trillion years to play in, "slow" can still happen. Given time enough, a black hole will completely evaporate through Hawking radiation.

A stellar mass black hole has a minimum mass of about three times that of the Sun. Pinging away particles one by one, it takes a long time to slog through six octillion tons of black hole: about 10^{66} years. To us, today, that seems like forever. But even that is the blink of an eye compared to the time it takes a supermassive black hole to blow away. The billion-solar-mass black hole that was once the Milky Way Galaxy (and Andromeda and several others from the Local Group) will take a whopping *10^{92} years* to evaporate to nothing.

And that's it. I'm out of analogies. I give up. I was hoping to come up with something like, if the life span of the Universe up until now were a single beat of a hummingbird's wing, then 10^{92} years would be like, well, like something that takes a *really* long time. But even comparing a single flap of a hummingbird's wing to the current age of the Universe falls completely and hopelessly short of comparing the present age of the Universe to 10^{92} years. That's just too long a time span. It crushes our sense of reality to dust. The closest analogy I could think of is to compare the mass of a proton to the mass of the entire Universe, but this analogy is useless. Analogies are supposed to make things easier to grasp, and who can grasp the mass of the proton, the mass of the whole cosmos, and then take the ratio?

Worse, *the analogy actually falls short of reality.* The ratio of 10^{92} years to the current age of the Universe is about 10^{82}, while the ratio of the mass of the Universe to the mass of a proton is 10^{79}. The analogy fails by a factor of 1,000.

So I give up. You're on your own for analogies now.

But perhaps we're done anyway. The most massive object in the Universe has evaporated away using the slowest process in the Universe. When it's done, there's not much left. The entire observable Universe will be only a million or two light-years across, and it will consist of countless electrons, positrons, neutrinos, a handful of exotic particles, and extremely low-energy photons. It will be an incredibly thin vacuum, far more rarefied than anything that exists today.

And that's it. That's all there is. Once the black hole is gone, everything familiar in our Universe will go with it.

The Universe will be dead.

THE DARK ERA: T + 10^{92}–∞ YEARS

The endless gulf of time stretches ahead of us now. At this point, our math breaks apart. The Universe is such a thin soup that it could be countless years before any two particles approach each other. And if they do, what will happen? If the two particles are both electrons, they repel each other and off they go in opposite directions. If one is an electron and the other a positron, they'll attract each other, collide, and poof! They'll make a pair of gamma rays that fly away.

But where will they go?

Every trace of the Universe we know today will be gone. No stars, no planets, no people. Not even matter. It will all have decayed away, eroded into an ethereally thin slurry.

10^{100} years, $10^{1,000}$, $10^{1,000,000}$. It's all the same. Nothing ever happens, and nothing ever will. The Universe is dark, randomized, silent. And it will remain so forever.

REBIRTH?

Oh, but there's that word again. *Forever.*

As we've seen many times in this chapter, nothing is forever. Maybe not even Universal death.

There are some faint hopes for the ultimate fate of the Universe. Most involve the complete destruction of the Universe as we know it and the reconstruction of something entirely new and different.

You might consider that a drawback.

But the alternative is the boring Universe where nothing ever happens. So let's see what we've got.

The Big Bang was a singular happening. Somehow, all matter, energy, space, and time were generated in that one event, forming the Universe as we know it.

But where did that event come from?

As discussed earlier, there are some theories that there is a meta-Universe, someplace *other* that exists outside of our framework of space and time. It developed a little quantum hernia, and this formed our own Universe. If that's really the case, then the death of our Universe isn't that big a deal in context. The other Universe may still be there, and it's possible that it budded off countless other universes too. These may all have vastly different laws of physics (maybe in one the speed of light is a few miles per hour instead of 186,000 miles per second, and in another the electron has more mass than the proton, instead of the other way around as in ours). It's also possible that our own Universe is doing this all the time—even now, tiny offspring universes are popping into existence in other "places," outside what we can see and investigate. However, according to everything we understand about physics, we can never physically learn anything about these other universes, so for all practical purposes they don't exist.

Of course, it's conceivable that in the next, oh, quintillion years or so our understanding of physics may change. I'll readily grant that! But for now there's not much we can say about this.

But maybe we're starting off on the wrong premise. Maybe we should ask: *was* the Big Bang actually the first cause? Or was there some other event that jump-started the cosmos?

There is another idea, still in its infancy, called the *ekpyrotic universe* (Greek for "from [or out of] fire"). According to this idea, the Universe is already incredibly old. At first, it was basically a giant void, with nothing interesting happening (much like the way we left it after 10^{100} years). According to this theory, there exist other universes with characteristics similar to those of ours now, but these other universes are outside our view. They exist in eleven dimensions instead of the four with which we are familiar (the three dimensions of space and one of time), and they float around in this extradimensional space. Called *branes,* short for *membranes,* these are all self-contained universes like ours in many ways, generally minding their own business.

But sometimes they collide.

You can picture these universes as parallel plates floating around. When they smack into each other, they shake up the contents rather vigorously. The theory predicts that the universes would get violently disturbed, with energy and matter being heated up tremendously, and space itself set to expand.

Sound familiar?

This may sound a little like fantasy, but it's all part of a set of very complicated but scientifically based math and physics theories. No one has any idea if these theories really are viable alternatives to the Big Bang model, or if it's just so much fantasy. But the ideas are internally consistent, and are being studied very seriously.

If they pan out, then there is some hope for the Universe, or at least for some meta-Universe: it means other universes exist, and they might be habitable. We can never reach them, so that's too bad for us, but maybe other species in those universes can survive.

And there is some hope for us as well. What happens once can happen again, especially if you wait long enough.

In 10^{100} or $10^{1,000}$ years, however long it takes, another brane may collide with ours. When it does, it may spark a reignition of the Big Bang, kick-starting our Universe once again. When that happens, pretty much everything that happened in the Universe before that point will be destroyed: kindling, if you will, for the fires of a new Universe. Again, that's too bad for us, but it does mean that there is a cyclical nature to reality, and a chance for life to rise anew.

Again, cold comfort. But it's a possibility.

DOWN THE LONG STAIRCASE

Yet another fate may wait in store for us in this far distant future, and we have even less of an idea of what will exist after it unfolds.

Objects have what is called an *energy state.* It's a bit like climbing a set of stairs. On the bottom step you are at the lowest energy state, and at the top you're at the highest. There are energy states in between too. It takes (muscle) energy to move up to higher states, and you give up that energy when you go back down.* Sitting at the bottom, you're as low as you can go, and you stop.

Atoms behave this way: electrons zipping around an atomic nucleus have certain energy states available to them, with none in between (just as you can't stand on the four-and-a-halfth step; it doesn't exist, so you

---

* This released energy mostly gets converted into sound (footsteps) and motion (momentum as you travel down).

have to be on either the fourth or the fifth step). This is one of the most basic ideas of quantum mechanics.

Perhaps the Universe behaves this way as well. We think of the vacuum of space as being, well, a *vacuum.* Empty. Devoid of matter and energy, and therefore at its lowest energy state.

But this may not be the case: we know that space bubbles and boils with energy at extremely small size scales (this is the basis for Hawking radiation, in fact). So what if we're not at the lowest energy level, the lowest energy state? What we're experiencing now would then be a "false vacuum state," and we might take that final step down, dropping to a lower state.

The starting point for this drop is difficult to predict. Maybe it's a quantum effect, again like Hawking radiation: somewhere, someplace in the Universe, a teeny-tiny bit of the Universe suddenly drops to the lower state. According to the theory, this one event acts as a trigger, pushing regions around it to drop into the lower state as well (imagine standing on the second-to-last step from the bottom with ten other people; you jump down and drag everyone else with you). A cascade starts, with more and more bits of the Universe dropping to the lower state.

This tunneling event, as it's called, would expand outward in a sphere at very nearly if not at the speed of light. It's a bit like a sugar crystal growing in a supersaturated solution; once you start it someplace, the other sugar molecules attach themselves there, growing rock candy.

In this case, though, the rock candy is the collapse to the true vacuum, and the sugar molecules are actually the fabric of space itself. The damage wrought is literally total.

Inside this expanding bubble of vacuum collapse, *the laws of the Universe change.* Space and time themselves are rewoven, becoming something entirely new, something the nature of which we cannot even begin to guess. Anything caught in this wave will be utterly destroyed.

And there is literally nothing to stop it. The entire Universe sits on the second-to-lowest state, so once poked by the expanding bubble, *everything* will collapse. Every star, every planet, every black hole, every human.

That would be pretty bad, were it to happen today. Odds are it won't; the chances of this happening are extremely small. But if this event were to wait, say, 10^{200} years, would that be so bad? I argue it would be *good.* By that time the Universe will be dead, stagnant, with nothing to show for all those years of activity. A collapse of the false vacuum to the true vacuum would possibly reenergize the Universe, giving it a second chance for life.

So there is some hope. You and I and even the entire Universe as we know it won't be there to witness it, much less survive it.

But afterward, a new Universe will be created, sparkling and clean and ready for a new start. In this case—and also that of the ekpyrotic universe, and maybe even other processes we haven't even yet conceived of—there is a chance that instead of a bleak and dim future, filled with nothingness for all eternity, there will be a rebirth of the Universe, and the rebirth of possibilities.

And if it happens once, it might happen again in another 10^{200} years, or $10^{1,000}$. And again, and *again.* Endlessly.

Rather than dealing out death and mayhem, destruction and chaos, the Universe will cyclically clean itself out, reboot, and set everything in motion once again.

Each time, perhaps, the laws will be different, and the characteristics of that future infinite parade of universes will be grandly set apart from what we know today. And despite our prejudices, the Universe appears to have no set of rules on how things need to be for complex chemistry, for life, to arise.

We don't know for sure if there are aliens in our own Universe now, though the odds favor such a possibility: there are 200 billion stars in the galaxy, and hundreds of billions of galaxies in the Universe.

And so I wonder: can we now multiply those odds by the number of potential universes that lie ahead as well?

If that's the case, then the Universe provides us a near-infinite number of do-overs, something I find very uplifting. It may seem that the Universe spends all its time trying to kill us, but in the end—the *very* end—there may yet be Life from the Skies.

EPILOGUE

What, Me Worry?

STILL HERE?

Good. It's been quite a ride, but I hope that while you read this book you weren't vaporized, crushed, irradiated, flung out into deep space, spaghettified, or had any of your protons decay.

We've covered a lot of scary ground (not to mention space and time). It seems as if the whole cosmos is trying to snuff us out. In a sense, it is—there's danger aplenty in the Universe—but we have to take a practical view here. We have to appreciate the vastness of space and time, and our ability to manipulate events around us.

Asteroid impacts provide an excellent example of practical versus theoretical danger. They have done ferocious damage to the Earth in the past, and our relatively fragile economic system could be destroyed by far smaller impacts than the one that did in the dinosaurs. To understand the actual danger, we have to balance the idea that they don't happen very often with the knowledge that they do in fact *sometimes* happen.

In your daily life, this may not present much of a problem (except for when you lie awake at night and your brain, unfettered by the common sense available during the day, is wondering if tonight is *the night*). But in the case of impacts, we can actually prevent them from happening. It would cost a lot of money (hundreds of millions to test various ideas, and hundreds of millions more to implement them), time, and effort. Can we afford to start worrying about this now? Can we afford to wait?

Scientists are asking these questions, because they have to ask Congress for a lot of money to be able to do anything about them. And what Congress decides influences your money (other governments may become involved as well). To make sure we get the right answers, we really have to understand the issues. Astronomers are looking at a theoretical danger and finding a practical action to take against it.

I hope this book has cleared some of that up. I studied gamma-ray bursts for many years, and I personally am not at all worried about them. Nor do I fret over the death of the Sun, the eventual decay of the Universe, or a black hole slipping through the solar system and snacking on the odd planet or two. That's because I understand the odds of these events actually taking place, and they are vanishingly small. There's no need to worry about them, whereas an asteroid strike or a particularly nasty solar coronal mass ejection can do a lot of damage. Even then we have it within our power to minimize their impact.

In an effort to make all this a little easier to digest, here is a table that gives the odds, the potential damage, and our ability to prevent the disasters described in this book. What you'll see is that chapters 1 and 2—asteroid impacts and solar events—cover the only two events we can do anything about. While they may not happen tomorrow, they will happen, and it's in our best long-run interest to do something about them.

Following the table is a description of how I came to those conclusions. Bear in mind that astronomy is a field of science, and that means that things change as better data and better ideas come along. Don't assume any of this is written in stone.

Of course, in 10^{40} years or so, even stone will be long gone.

EVENT	DAMAGE	ODDS OF FATALITY (PER LIFETIME)	PREVENTABLE?
Asteroid impact	Local for a small rock, global for a big one	1 in 700,000	Almost 100% preventable Identify potential impactors, then blow them up or push them out of the way
Solar flare/CME	Collapse of power grid, potential ozone depletion	0*	Not preventable, but mitigable Build robust power grids
Supernova	Ozone depletion, radiation	1 in 10,000,000	Not preventable
Gamma-ray burst	Ozone depletion, radiation, setting planet on fire	1 in 14,000,000	Not preventable
Black hole	Destruction of Earth	1 in 1,000,000,000,000	Not preventable
Alien attack	Humanity wiped out by aliens; space bugs give us runny noses	?	Preventable, assuming we colonize the galaxy first; otherwise, forget it
Death of the Sun	Earth cooked to a crisp	0†	Not preventable, but we have a long time to go yet
Galactic doom	Ice ages, radiation, eaten by supermassive black hole	0†	Not preventable, but again, none of these will happen on a human time scale
Death of the Universe	Decay of all matter, collapse of false vacuum	0†	Not preventable, but dwarfs any time scale we can imagine

* Fatalities are very unlikely from a solar event, but they can still cause extensive damage.

† These events all take billions of years (at least!) to unfold, so the chances of their happening during your lifetime are zero, but they are inevitable over longer times.

ASTEROID AND COMET IMPACTS

Of all the woes facing us from space, this is the one that is nearly 100 percent preventable. Scientists and engineers have viable ideas on how to stop big impacts—ones big enough to do significant damage. The real problem is in finding these objects in time, and even that is improving as more surveys of the sky find more objects. However, it's physically impossible to find every single potential impactor; some come from so far out in the solar system that we simply cannot see them until they are on the way here.

So no matter how many we find, there is still some risk. The American astronomer Alan Harris has composed a table of risks from impacts, and the results are surprising: if you live in the United States, the overall risk of dying from an impact in your lifetime is only 1 in 700,000, somewhat less than being killed in a fireworks accident, but still more probable than being killed on an amusement park ride or by an act of terrorism.

Interestingly, before we started surveying the skies, those odds were calculated to be about 1 in 70,000, which was ten times higher. Why? The old numbers were based on statistics using previous impacts as a basis, and the statistics on those are a bit spotty. Now that we have surveys peering at the skies, most of the big impactors have been found. Since these tend to kill lots of people (mass extinctions, anyone?), eliminating them from the known list of potential impactors dropped the odds of getting killed significantly. Actually, the odds haven't really changed; we just know how to calculate them better. It's like the difference between trying to figure out your odds of winning a poker hand before and after the cards are dealt. Once you see those cards, your odds are far easier to calculate.

So how's that for a return on investment? This makes it very clear why just getting out and finding these things is so important. We spend billions on terrorism, but the risk from an asteroid impact is actually *higher.*

Mind you, this does not include comets, which can come from deep space and have orbits that are not easy to determine in advance.

SOLAR EVENTS

These are tricky things to evaluate. For one thing, big solar blasts are rare, making statistics spotty. Also, the effect they have on Earth depends on many factors, including time of year: during the peak summer and winter months the power grid on Earth is heavily loaded with electricity and more susceptible to flares and CMEs, while in the spring and fall months the grids are more robust.

Perhaps the most important issue is that few or no deaths will result directly from a solar event. There are some direct effects like exposure to radiation for long-duration or polar airline flights, but these are difficult to pin down. Astronauts are at risk, but none has ever been killed by a solar event, though this will certainly be a problem for long-duration lunar and Martian voyages.

The large effects are all indirect: loss of power, loss of communication, and so on. These can cause deaths, of course: heatstroke in summer, hypothermia in winter. But the relationship is difficult to determine.

While a whopping big solar event can seriously impair or destroy a nation's infrastructure and economy, in general it will not directly cause deaths. So we have to rate this a zero for human fatality, but with an asterisk as a nod to the destructive power it has in other ways.

SUPERNOVAE AND GAMMA-RAY BURSTS

Supernovae happen about once per century in any given galaxy. But galaxies are huge, and the damage from an exploding star is limited by proximity: it needs to go off less than about 25 light-years away to impart significant damage to Earth's ozone layer. This only happens about once every 700 million years or so. Assuming the event would cause a mass extinction, killing everyone on Earth, the odds of your specifically dying from one over your lifetime are therefore about 1 in 10 million.

Gamma-ray bursts are a little different: they inflict far more damage, and are dangerous from distances of more than 7,000 light-years. But they are also rarer than regular supernovae and also pickier about targets: we have to be in the path of the relatively narrow beam to get hurt. All in all, these effects cancel out, leaving us with just slightly lower odds as being killed by a supernova: for GRBs, the odds are 1 in 14 million.

In both cases, you're literally more likely to be killed by a shark.

BLACK HOLES

The odds that a black hole will get close enough to the Sun to do any real damage are very low. A normal star passes within about three light-years of the Sun about every 100,000 years, and there are something like 20,000 normal stars in the Milky Way for every black hole (assuming 200 billion stars and 10 million black holes, the latter of which is probably also an overestimate). This means a black hole gets within three light-years of the Sun, statistically speaking, every two billion years, or three times over the solar system's current lifetime.

From that distance, the black hole can't do very much to us; remember, from a long way off its gravity is no different from that of a normal object. Even a black hole with ten times the Sun's mass won't do much from that distance.

If the black hole is actively eating, then it will emit X-rays that can hurt us from a greater distance. But three light-years is probably a safe distance; from that far away the X-ray emission powered by a black hole is actually far less than the X-rays you'd expect from a solar flare. Even a closer passage by about a light-year would only cause minimal damage to the ozone layer, and that kind of close shave would be rarer still, both because the odds of one getting that close are small and because active black holes are far less common than quiescent ones.

To be as bright as a big solar flare in X-rays, a typical black hole would have to get within about 150 billion miles of the Earth. This is an extremely unlikely event, happening about once every 100 trillion years. Obviously, this is not likely to have ever happened in the history of the solar system, nor is it ever likely to.

So, your lifetime odds of Death by Black Hole are about one in a trillion.

ALIEN ATTACK

So how do you calculate the odds of being attacked by aliens? We could use the Drake Equation outlined in chapter 6, but as we saw, the output of that equation (the number of advanced civilizations in the galaxy) is not well constrained, as scientists say: it could be one (us), or it could be millions.

Even if the galaxy is buzzing with life, it's nearly impossible to quantify how many of these aliens would be hostile, and how many would come here specifically to wipe us out. It only takes one race, of course, but what are the odds?

I argue that we can't know. The best we can do is say that if a race wants to wipe out all life, *everywhere,* then they must visit every planet not only with life on it but also *capable* of having life on it. Why take chances?

We have enjoyed a nearly uninterrupted existence of life on Earth for over three billion years. During that time, we have not seen a single alien intelligence sterilize the planet. That puts a lower limit on the odds of alien-induced extinction to one in three billion. Given that a species can explore the entire galaxy in a small amount of time (a few million years) compared to the amount of time life has existed (billions of years), the odds are for all practical purposes zero.

As I pointed out in the chapter, it's fun to think about, and it makes a great bedtime story, but in the real world I think estimating the probability as equal to zero is close enough.

Also, we cannot really assign odds to a space virus or bacterium turning us into goo anytime soon either. We have not seen a single example of such a beast, though of course we haven't been looking all that long. Still, because of the lack of data, we have a true unknown here. Personally, using just my guts and hunches, I would put the odds in the range of billions-to-one against, but that is not very scientific. So to be truly skeptical, as any real scientist is, I will have to leave this blank, and hope that advances in astrobiology will allow us to make some safe estimates of the odds sometime soon.

DEATH OF THE SUN

Of all the ways the Cosmic Grim Reaper can pay us a visit, just this one and one other (Death by End of the Universe) have odds of 100 percent. You just have to wait long enough!

The Sun will begin to die in six billion years, and unless we develop technology indistinguishable from magic, there's nothing we can do about it (short of migrating to another star). But this is a funny case: the odds of it happening in your lifetime are exactly zero, but if you live for six billion years, the odds jump to 100 percent. However, this being an almost entirely predictable event, I'm going to have to go with the lower-limit case.

GALACTIC DOOM

Just living in the Milky Way Galaxy is dangerous, but *how* dangerous? Actually, it's not hard to get some statistics. For example, magnetars—supermagnetized neutron stars—can cause damage from several thousands of light-years away. Like GRBs they are created in supernova explosions, and are also quite rare. It's reasonable then to assign the same death-inducing stats to them as for GRBs, or about 1 in 10–20 million or so. It's worth remembering that no magnetars are known close enough to hurt us, and it's likely that if any were that close we'd see them.

Plowing into a dense dust cloud is a rare event as well, happening on the order of once every billion years. This happens most often when the Sun enters a spiral arm of the galaxy. The Sun is located about 6,400 light-years from the nearest spiral arm, and so even if we were headed straight at it, it would be another 10 million years or so before we hit it. Therefore the odds of hitting a dense dust cloud are extremely low, but go up substantially in a few million years. Statistically speaking then, the odds of dying from entering a dust cloud are essentially zero right now, and will be for quite some time.

The same is true for solar oscillations out of the plane of the Milky Way; we are currently heading up into the danger zone, but won't be there for at least another 20 million years. So again, currently the chance of getting killed by a flood of intergalactic cosmic rays is zero.

Finally, the collision with the Andromeda galaxy has several avenues for killing us: we could get dropped into the galactic center whereupon we get eaten by the supermassive black hole dwelling there, or we could get close enough that high-energy radiation from said black hole may do us in, or the (possible) vast amounts of new stars born in a cosmic baby boom could produce a nearby supernova (making the "boom" particularly appropriate). While any of these are dangerous, this collision is another event that cannot happen for several billion more years, so the odds of its killing you are currently zero.

DEATH OF THE COSMOS

Like the Death by Solar Death, this is inevitable. Take your pick: do you care more about the Milky Way colliding with the Andromeda galaxy (which will happen before the Sun dies), or the eventual decay of protons, or the evaporation of black holes, or the total annihilation of the Universe by quantum collapse to the true vacuum state?

But the time scales! Billions of years, nonillions of years, vigintillions* of years . . . if you wait long enough, any number of bad things

---

* Yes, I had to look up those two words.

will happen, and any one of them is pretty final. Again, though, like the death of the Sun, most of these events are not random, but happen at a certain pace. Protons almost certainly will decay, but not for more than 10^{33} years at least. So we're safe over our lifetimes from these disasters.

There is some finite chance at any one time that we'll collapse to the true vacuum state, but since this is entirely theoretical, I don't think any number has ever been assigned to it. The odds are so low that even unlikely events—like getting eaten by a rogue black hole—are fantastically more likely to happen long before the Universe decides to erase all of space and time. I would hang a big zero on this one as well.

LIFE, THE UNIVERSE, AND EVERYTHING

The last few events are interesting: they won't happen in your lifetime, but they *will* eventually happen. This skews the statistics, since calculating a lifetime risk from such an event doesn't really work. The death of the Sun won't kill you—unless you plan to see the ripe old age of six billion—but it's out there, someday.

This gives us some insight on how much we should worry about these astronomical harbingers of doom. The only two we *can* do anything about (asteroids and solar events) are things we *should* do something about. The cost isn't that big a deal in the long run, and the savings are enormous.

As for the rest, well, you shouldn't fret too much. They are certainly fun to think about, and by studying them we learn more about the awesome nature of the Universe, its scale and its capabilities. One of the biggest thrills of science, and its ultimate goal, is *understanding.* Maybe the price to pay for that is a little bit of fear, but I think in this case it's only a very tiny price. A bargain, in fact.

Sure, the Universe is scary. But it's also beautiful. A supernova can kill us, but the expanding wave of gas creates one of the most intricate and delicate objects we can see. Colliding galaxies dance to a tune millions of years long, creating graceful and elegant structures for us to ponder. The Sun may evolve into a red giant, but the planetary nebula it may eventually become will shine green, red, and blue, and will grace the skies of countless planets for a dozen millennia.

And there's more to this story as well. Supernovae create heavy elements and then seed nearby gas clouds with them. These elements are necessary for planets to form and for the rise of life. Colliding galaxies make new stars, new chances for life. Even the death of the Sun means its material is returned to space where it may get used again. Life, death, and life: this is the real story of the Universe.

Most of the Universe is lethal, but our little section of it is pretty cozy. The cosmos takes away, but it also provides. So go outside, enjoy a sunny day or a star-filled night. There is danger to avoid, but also beauty to behold, and you understand a little bit more about that sky now.

And understanding is always good.

Acknowledgments

A book that deals with everything from asteroid impacts to the quantum decay of the Universe is something of a big task, and not one to be undertaken either lightly or alone. If you took everyone I need to thank and put them in a room together, they would collapse into their own event horizon.

But let's allow a little quantum Hawking radiation to escape, and see who pops up.

First, thanks go to my agent and to my editor, Loretta Barrett and Alessandra Lusardi. Loretta is a good person to have on your side in negotiations, and when the dust settled Alessandra did a fine job encouraging me when I suddenly realized I was actually going to have to write a book. Also my thanks to the folks at the Boulder Conference on World Affairs; it was because of that fine event that I met Loretta. I liked it so much I moved to Boulder.

I have a decent grasp of astronomy, but not anywhere near the level needed to cover the ground of this book. So I want to toss rose petals at the feet of the scientists who helped, tore apart, edited, dissected, examined, corrected, folded, spun, mutilated, and finally tech-edited what I wrote. They include Dan Durda, Ed Lu, Rusty Schweickart, Al Harris,

Craig DeForest, Barbara Thompson, Caspar Ammann, Christopher Balch, Neil Gehrels (who was running an entire NASA mission during the time he took to help me!), Andrew Hamilton, Seth Shostak, Harriet Hall, Rich Pogge, Michelle Thaller, and Fred Adams. Any mistakes in this book are no doubt due to my own transcription issues after hours on the phone with all these fine folks.

My appreciation also to the people who personally gave me artwork for the book: Dan Durda, Dana Berry (his stuff is *amazing*), David Hardy, Chris Setter, and Aurore Simonnet.

To my friends, my co-workers (especially Sarah Silva), and my boss, Lynn Cominsky, who put up with me during this turbulent ride: thanks. A very special head bob to Fraser Cain of UniverseToday.com, who helped immensely and gave me tons of good ideas to help support the book online.

Thanks go to the BABloggees and other minions who tolerated fewer blog posts for months while I scratched this thing out.

My family: again, you've been great during this ridiculous time. For my mom especially, who has encouraged me constantly, and very much for my dad, who would have loved to read this book.

And, as usual, I owe everything to Marcella and Zoe. I don't believe in the supernatural of any kind, but if there is some sort of imprint, any echo of personality that embeds itself in our matter, then my love for both of you will endure for about 10^{40} years. It would take me that long to express what it means to me anyway.

Appendix

Nearby Stars (Distance < 1,000 Light-Years) That Will Eventually Go Supernova

PROPER NAME	GREEK LETTER DESIGNATION	CONSTEL-LATION	MASS (SOLAR MASSES)	DISTANCE (LIGHT-YEARS)	APPARENT MAGNITUDE
Spica	Alpha Vir	Virgo	11	260	1.04
Shaula	Lambda Sco	Scorpius	11	365	1.63
Dschubba	Delta Sco	Scorpius	12	400	2.32
Al Niyat (Tau)	Tau Sco	Scorpius	12	400	2.82
Betelgeuse	Alpha Ori	Orion	12 to 17	425	0.7
Adhara	Epsilon CMa	Canis Major	10 to 12	430	1.5
-	Kappa Sco	Scorpius	10.5	450	2.41
-	Zeta Oph	Ophiuchus	20	460	2.56
-	Pi Sco	Scorpius	11	500	2.89
-	Epsilon Per	Perseus	14	500	2.9
Mirzam	Beta CMa	Canis Major	15	500	1.98
Al Niyat (Sigma)	Sigma Sco	Scorpius	12 to 20	520	2.91
Graffias	Beta Sco	Scorpius	multiple; at least two are 10 solar masses	530	2.5

PROPER NAME	GREEK LETTER DESIGNATION	CONSTEL-LATION	MASS (SOLAR MASSES)	DISTANCE (LIGHT-YEARS)	APPARENT MAGNITUDE
Antares	Alpha Sco	Scorpius	15 to 18	600	0.96
-	Gamma Cas	Cassiopeia	15	610	2.47
Enif	Epsilon Peg	Pegasus	10	670	2.39
Saiph	Kappa Ori	Orion	15 to 17	720	2.06
Rigel	Beta Ori	Orion	17	775	0.12
Alnitak	Zeta Ori	Orion	binary; 20 and 14	815	1.74
Alfirk	Beta Cep	Cepheus	12(?)	820	3.23
Saif al Jabbar	Eta Ori	Orion	binary; 15 and 9	900	3.35
Mintaka	Delta Ori	Orion	20+	915	2.23
-	Zeta Per	Perseus	19	1,000	2.84
Meissa	Lambda Ori	Orion	25	1,000	3.39

Notes: Several of these stars are binary, and the component masses are listed. Masses for many are approximations based on the type of star and the known brightness. Apparent magnitude is a brightness system used by astronomers: larger numbers mean fainter stars, with the typical naked-eye limit about 6; all the stars listed can be seen without optical aid. None of them is close enough to directly damage the Earth when they explode.

Index

Page numbers in *italics* refer to illustrations.